בס"ד

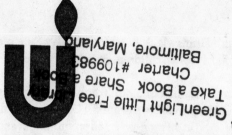

**SHALOM
TORAH
CENTERS**

מוסדות שלום לחינוך ילדים

Shalom Torah Centers: So Near and Yet So Far

Just across the Hudson River from the dynamic Jewish centers of New York City lies the State of New Jersey. Except for a few enclaves of strong Jewish awareness, crossing this river is like crossing into another world. There are more than a half million Jewish people in New Jersey, untold thousands of them children. Most of these children are mainstreamed into the public education system, their Jewish identity sublimated in the quest for "acceptance." Blended into this system, they are never given the opportunity to discover the singular beauty of our Torah, its heritage and traditions.

At worst, these children will intermarry and be forever lost to our people. At best, they will carry their Jewishness through life as an unwelcome burden, and their own children will be that much further from the rest of us.

It is a tragedy in the making, a tragedy that need not be.

In a courageous effort, a group of Torah scholars from Beth Medrash Govoha in Lakewood, New Jersey, joined together in the early seventies to create the Shalom Torah Centers. Although committed to spreading the light of Torah to all the Jewish people of New Jersey, Shalom Torah Centers were clearly focused on the children.

The first Shalom Torah Center was established in Manalapan Township in 1973 with an enrollment of thirty-one children. Viewed from the perspective of the overall situation, it was a small nick in a mountainous problem, but it was a beginning nonetheless. Viewed from an individual perspective, however, it represented thirty-one victories, for every Jewish child is priceless in its own right. Every Jewish child carries with it the prayers and sacrifices of countless generations of its forebears and the hope for all future generations.

Looking back, the first eighteen years have been a time of growth and consolidation for Shalom Torah Centers. That first small school has grown into a blossoming network of four Talmud Torahs (Manalapan,

East Windsor, Matawan, South Brunswick), two Hebrew Day Schools that collectively reach hundreds of Jewish children, and a highly effective Adult Education program.

In Twin Rivers, the Morris Namias Shalom Torah Academy has transformed an entire community, providing a beautiful Orthodox Shul, excellent education for the children in a Torah-oriented environment and a full program of adult education courses. The Shalom Torah Academy of Englishtown-Old Bridge has also experienced phenomenal growth. It, too, has assumed the role of a much needed center for both children's and adult education.

It has been the experience of organizations involved in this kind of work that most people drift away from the Torah and traditional Jewish values only because of ignorance. They need only be reached and they will respond. At Shalom Torah Centers, this has been found especially true with the children. Whether at a Shabbaton weekend or during the daily school curriculum, it can be seen in the profound and genuine delight that lights up their eyes as they discover the heritage that is rightfully theirs.

Successfully reaching the children has often resulted in transforming the entire household. Parents, impressed by the positive influences of Jewish education on their children are drawn to attend adult education classes and the innovative "Chavrusa programs" also offered by Shalom Torah Centers. Currently, close to 100 members, men and women of the Shalom communities, meet on a weekly basis with scholars of the Lakewood community to study Torah and learn about Yiddishkeit on a one to one basis. This dynamic program alone has borne much fruit as many of the participants and their families have become fully observant.

As Shalom Torah Centers face the future, they can point proudly to their record of achievements. They have emerged as a major force in the battle against the tide of assimilation in America. The present network of schools continues to blossom; the quality of the education in both Torah related and secular subjects is uniformly excellent; parent and community response is enthusiastic. Indeed, a strong nucleus has been created, yet for every child and adult reached by Shalom Torah Centers, there are hundreds more to be reached in communities across New Jersey and its neighboring states.

Yes, the future unfolds with great promise, provided that there is a major investment of long hours, hard work, and new resources.

So near and yet so far.

THE OBVIOUS PROOF

A Presentation of the Classic Proof of Universal Design

GERSHON ROBINSON
MORDECHAI STEINMAN

CIS PUBLISHERS

New York · London · Jerusalem

Published and distributed
in the U.S., Canada and overseas by
C.I.S. Publishers and Distributors
180 Park Avenue, Lakewood, New Jersey 08701
(908) 905-3000 Fax: (908) 367-6666

Distributed in Israel by
C.I.S. International (Israel)
Rechov Mishkalov 18
Har Nof, Jerusalem
Tel: 02-518-935

Distributed in the U.K. and Europe by
C.I.S. International (U.K.)
89 Craven Park Road
London N15 6AH, England
Tel: 81-809-3723

Book and cover design: Chaya Bleier, Deenee Cohen
Typography: Chaya Bleier
Cover illustration: Tova Leff

ISBN 1-56062-175-3 hard cover
1-56062-183-4 soft cover

Library of Congress Catalog Card Number:
92-076188

PRINTED IN THE UNITED STATES OF AMERICA

Dedicated in the memory of

R'Moshe Aharon Hirschfeld צ"ל

TABLE OF CONTENTS

INTRODUCTION

W hat would you think if you found a watch in the desert? Would you think it had come there of itself? Or would you think someone made it and brought it there? And, if so, does it not follow that the world, which is infinitely more complex than a watch, also has a maker?

This hypothetical situation forms the basis of the oldest and most classic proof of universal design and the existence of a Creator. In spite of the seemingly overwhelming force of the "watch in the desert" argument, however, denial of the Creator still persists in secular intellectual circles. The purpose of this book, in addition to a new and contemporary presentation of the classic proof of universal design, is to expose and examine the psychological mechanisms that trigger this irrational denial.

If we take the long historical view, we will find that people have always believed in a deity as a matter of course, this being the only reasonable explanation for how the universe in all its complexity came into being in the first place. Therefore, no matter how much time was devoted to making a living and obtaining material things, people always "left room" in their lives for G-d.

In modern times, however, with the phenomenal advances in science and technology, many people have chosen thoroughly materialistic lifestyles which eliminate the factor of G-d completely; perhaps as man continues to succeed in mastering his environment, he finds it more difficult to accept the existence of a superior omnipotent Master. The concept of G-d is an "invention of man," the materialists argue, a vestige of the primitive past when man needed a G-d concept to help him cope with the vicissitudes of life. Previous generations would have thought differently, the materialists argue, if they had known what is known today.

But how does one explain the origin of the universe and its unfathomable complexity? A thorny question, indeed, but modern science has come to the rescue with assorted G-dless theories and philosophies that raise more questions than they answer. Thus, the existence of G-d, which used to be assumed as a matter of course, has suddenly become a subject of debate.

In the following pages, the movement away from G-d and toward atheistic materialism will be examined, and the flimsiness of its scientific foundations will be exposed. Some of the most glaring flaws are actually discussed in the writings of modern scientists themselves, but overall, there seems to be a peculiar myopia with regard to this question, a myopia that will be analyzed and hopefully explained.

The thesis presented in this book is derived from solid

psychological principles, and its application to the issues at hand is reinforced by a rather new and innovative observation that convincingly drives the point home.

The Obvious Proof is a revised and expanded edition of the previously self-published and highly popular *The 2001 Principle*. We wish to thank C.I.S. Publishers for the devoted efforts that have gone into the production of this new edition, with a particular note of acknowledgment to Chaya Bleier who was responsible for the design and typography of the entire book.

In closing, we wish to offer a prayer of gratitude to the Creator of the Universe for His infinite kindnesses and for allowing us the privilege and honor of glorifying His Name through our own humble efforts.

Gershon Robinson
Mordechai Steinman

CHAPTER ONE

COGNITIVE DISSONANCE: THE HUMAN TENDENCY

Imagine, after having recently bought an expensive clock or car, you see an advertisement extolling the virtues of the same item. If you are like most people, you will find yourself attentive, and you will consider the ad quite credible. On the other hand, if you would see an advertisement extolling the virtues of a competing model, you would experience some discomfort and an urge to discredit the message. These reactions are common and natural, because everyone feels at ease being right and dislikes being wrong. Being right induces a feeling of superiority, and being wrong induces a feeling of inferiority. Therefore, anything conveying the message that one is wrong is irritating and is met by discomfort; it poses a threat to one's self esteem. If a person who had made an incorrect choice accepts the new information, he will suffer a blow to his ego.

Psychologists often call this negative reaction "cognitive

dissonance," a sort of "static" produced in a person's psyche when he confronts information telling him he is wrong. A physical consequence of cognitive dissonance is an actual squirming of one's body to one degree or another. Everyone, young or old, is subject to this reaction when new information runs counter to his desires. Being confronted by criticism about something to which one feels attached, by a challenge to what one feels is true, inevitably produces cognitive dissonance and some degree of squirming.

It is important to realize that cognitive dissonance is created not only in the marketplace of products. Dissonance is created even by facts, ideas and situations, when one feels that the new information is not consistent with one's personal status quo. Anything new must "square" with one's present feelings and thoughts, where a person is now. Otherwise, "dissonance" arises. Whenever something comes along which does not square, cognitive dissonance automatically appears in the human subconscious.

We react to "dissonance" in two basic ways, though one reaction predominates. Sometimes, we accept the new information, face the uneasiness and adjust to it. In the case of the ad for the product we did not buy, we accept the possibility that the other object is, in fact, better. We keep an open mind and admit the possibility of error despite the resultant loss of self esteem.

In many cases, however, people avoid exposure to the new assertion entirely. Most people are not even able to listen to an advertisement praising a competing product, let alone believe it. Moreover, the greater the investment, the greater the dissonance. In effect, cognitive dissonance can completely override the human desire for truth. If one has "bought into something," if one has made a large enough investment in a certain product, belief or idea, then any assertion indicating that the investment was bad stands a good chance of being disregarded, even if it is true.

Social influences are a major factor in determining which ideas people consider to be long accepted and therefore adoptable. New ideas are viewed as too revolutionary for serious consideration. Another source of cognitive dissonance is information or situations that are complicated and hard to understand. When a person is confronted by new information beyond his powers of comprehension, dissonance arises, because the hard-to-understand data makes the person feel inferior.

For example, if after repeated readings about a new scientific finding, one cannot understand what the scientists found (or what the finding is), cognitive dissonance will be created, no less than when one is faced by the prospect of personal error. When a person is confronted by information which he simply does not understand, such as a new scientific finding, his ego is hurt. Because he cannot understand the information, he loses self esteem, just as with information which indicates he erred. In other words, just as information which indicates error creates cognitive dissonance, so does information which indicates lack of understanding. Rather than accept the information and suffer the "dissonance," people often disregard it, even if it is true.

For example, a recent publication of the Jewish National and University Library[1] commemorates the achievements of Albert Einstein and contains the following statement about the special theory of relativity, which Einstein proposed in 1905: "The rejection of the relativity paper was almost universal."

Without going into detail, the rejection was obviously not based upon facts, because today, the ideas of Einstein are fixed pillars upon which the whole of science rests. Moreover, Einstein revolutionized not only science but also philosophy and political relations. Indeed, in the words of a 1952 editorial,[2] Einstein "opened a new era in the history of mankind . . . he has impressed the stamp of his ideas on the whole of the twentieth century." His ideas would bring about atomic power, with all of its ramifications, fundamentally new insights into time, space, matter, gravity and energy and new approaches to scientific inquiry.

Nowadays, people are surprised when they learn that Einstein's ideas were originally discounted, almost by everyone. Yet this is generally the case with new ideas when the ideas are difficult to understand.

Take the idea of the earth being round, not flat. Put forth formally after direct observations through a new telescope, this idea was not open to any question whatsoever. There was clear proof. However, traditionally the earth was considered to be flat, and people wondered aloud, "If these observations are right, and the earth is round, why aren't we all falling off?" Rather than accept an idea they could not understand, in this case heretofore unknown principles of gravity, people rejected the new idea completely. Rather than admit that they lacked understanding, people turned their backs on the truth.

The social influences of which we spoke were also at work, keeping people from being open-minded. The general consensus held a certain opinion about the shape of the world.

That opinion was recorded as fact in all the textbooks. People had always assumed that the earth was flat. They had lived with that assumption and had built upon it. To face new, contradictory information would require a tremendous amount of readjustment and change. People avoid such things. Whenever a new idea conflicts with widespread belief and general public opinion, individuals suffer cognitive dissonance.

Conclusive proof that the earth was round could not be accepted, because the implications were much too painful. In order to avoid having to make two statements, "I do not understand," and "Everyone has been wrong," people completely ignored *hard facts*. For personal and social reasons, people would not and could not face the truth.

THE DISCOVERY BY EDWIN HUBBLE

Another example of this phenomenon concerns man's ideas about the size of the universe. Today, it is common knowledge that the Milky Way Galaxy, which contains the earth, the sun and some fifty billion stars, is but one of billions of galaxies, each containing billions of stars of its own. For a long time, however, scientists thought the only galaxy in the universe was ours, the Milky Way. Only in 1917 did clear proof against the one-galaxy view of the universe begin coming in. It was revealed to be wrong, small minded and completely void of true understanding. Because of cognitive dissonance, however, scientists literally "closed their eyes" to the new information for almost eight years.

In 1917, using photographs taken through telescopes

trained on the night sky, an astronomer named Ritchey found that the entities previously thought to be novae in nebulae *inside* the Milky Way are actually separate galaxies hundreds of thousands of light-years *beyond* the Milky Way. An astronomer named Curtis soon presented similar photographic evidence, and this began a fascinating sequence of events described in a popular publication:[3]

> Curtis undertook to announce to the world what seemed to him unequivocal evidence that nebulae containing faint novae are separate galaxies. But the world—at least of astronomy—was not yet ready to accept the huge universe Curtis had to offer. An historic wrangle ensued, continuing at one astronomers' conference after another from 1917 through 1924 ... Then, abruptly, at a conference session of January 1, 1925, the great debate ended with the reading of a communication from the California astronomer Edwin Hubble.
>
> Hubble himself was absent, and word from him was eagerly awaited because he was busy taking the first "look" at the sky with the brand-new 100-inch telescope at Mount Wilson—then the world's largest. The momentous news he had to report was, the new telescope had resolved images of stars in three so-called nebulae: M31 in Andromeda, NGC 68_{22} and M_{33}. What was more, some of the resolved stars were Cepheids, the marvelous beacons of Miss Leavitt, and they proved by their faintness and periods that all three nebulae really were galaxies far, far beyond the island universe containing man.
>
> Once the reality of galaxies was finally proved, the study of them shot forward as if it had been building up steam all

through the previous decades of doubt and debate. Assisted by Milton Humason, one of the most technically adept observers in all astronomy, Hubble proceeded with the one-hundred-inch telescope to scale the ramparts of heaven. From globular clusters a few light-years away he reached out to the limits of the Cepheid measuring rod, a sphere about three million light-years in radius that encompassed some twenty galaxies. From there, using as rough yardsticks the brilliant blue supergiants in the spiral arms of galaxies, he went on to chart a further sphere—containing two hundred more galaxies—which is now estimated to be some thirty million light-years in radius. Still farther out, where single stars were no longer visible, Hubble approximated distances by the average innate brilliance of whole galaxies—ones corresponding in type to those he had already studied in the thirty-million light-year sphere. With this new measure, he raced on out to the then visible limits of the universe, over a thousand million light-years away.

In the course of this stupendous intellectual voyage—a rolling-back of human horizons unparalleled in previous history and not likely to be equalled ever again—Hubble calculated there were almost as many galaxies outside the Milky Way as there are stars in it.

In short, it would be an understatement to say that the astronomers up until 1925 were in error about the universe. They had been fundamentally incorrect about its total nature. And resistance to the idea of a multi-galaxy universe arose precisely because, prior to Ritchey and Curtis, the one-galaxy view of the universe was almost universally subscribed to. As had happened centuries earlier when new evidence was discovered about the shape of the earth, the discovery of additional galaxies would require a sweeping overhaul of scientific thought. Textbooks would have to be rewritten. Many prior assumptions would have to be re-examined. For anyone, admission of a gross error is painful. Because the information that became available in 1917 demonstrated not only great error but also human puniness (the earth is but a "speck"), scientists stubbornly refused to accept the information for almost eight years.

To review, unequivocal evidence about the size of the universe was available for all to see in 1917. Nevertheless, the implications of this evidence were so startling, so unsettling, that the scientific mind "snapped." Cognitive dissonance was at work. Emotional, psychological and other pressures prevented scientists from accepting direct observation. Normally rational minds were suddenly irrational. Not until 1925 were they finally able to see reality clearly.

WHY EINSTEIN WAS INITIALLY IGNORED

Still another example. As we mentioned above, when Einstein presented his theory of relativity to the scientists of the world in 1905, he was met by rejection. Einstein's ideas challenged long-accepted "facts." As objects accelerate towards the speed of light, Einstein proposed, not only do they age more slowly, they also get heavier. These phenomena, now established facts, are not apparent at all to the man on the street, in a way not unlike the facts of the roundness of the earth and the existence of other galaxies. Thus, people cannot comprehend how such things could be. A personal bias exists—one which can affect even scientists. Therefore, in spite of their airtight logic, the ideas of Einstein at first were not believed.

Just as in the other examples given, the negative reaction to the relativity theory sprang from social and other reasons as well. According to a former president of the Royal Astronomical Society, Einstein's ideas, once established, represented "not the discovery of an outlying island but of a whole continent of new scientific ideas."[4] For many, the implications of Einstein's ideas were simply too much to handle. But this reaction is not really very surprising, once we remember the power of cognitive dissonance.

Einstein was revolutionary; he "battled against all previous hypotheses."[5] When he contradicted the accepted theories by suggesting that light, as matter, is divided into atoms, he was met by ridicule. But he did not care. His sole intellectual guidepost was reason, and he followed it wherever it took him. Indeed, a publication we have cited previously[6] records a description of Einstein by one of his colleagues as the epitome of logic and reason.

Throughout his life, Einstein believed human reason could lead to theories able to provide correct descriptions of physical phenomena. In building a theory, his approach had something in common with that of an artist; he would aim for simplicity and beauty (beauty for him was, after all, essentially simplicity). The critical question he would ask when weighing an element of theory was, "Is it reasonable?"

No matter how successful a theory appeared to be, if it seemed to him not to be reasonable, he was convinced that the theory could not provide a really fundamental understanding of nature.

History shows Einstein was unperturbed by the incompatibility of his ideas with those of his peers. He wrote simply, "It will be possible to decide whether or not the foundations of relativity correspond with fact only if a great variety of observations is at hand."[7] Eventually, the observations were made, and Einstein's impeccable logic was indisputably confirmed.

Ultimately, then, Einstein's greatness lay in his ability to rise above cognitive dissonance. Einstein achieved a level of detachment from personal, social and other biases that invariably prevented his fellow scientists from seeing reality. For Einstein, truth was what mattered. The implications were unimportant.

INTELLIGENCE IS NO IMMUNITY

Yet, history shows that, as exceedingly strange as it may seem, Einstein himself once succumbed to cognitive dissonance. For this, we must revisit the astronomer Edwin Hubble who, the reader may recall, discovered the decisive evidence about galaxies by means of direct telescopic observation. In 1925, besides having discovered conclusive proof that there were countless stars beyond the Milky Way, Hubble also had proof that these stars were continually moving away from earth, indicating that the universe was expanding:

> A decade before Hubble found the cosmos to be expanding, Einstein's equations showed it should be either expanding or contracting. *Unable to believe his own results, Einstein rewrote them to let it stay static.*[8]

In a letter,[9] Einstein wrote, "The circumstance of an expanding universe is irritating." In another letter,[10] he wrote, "To admit such possibilities seems senseless to me."

In sum, the existence of a round earth, of billions of galaxies other than ours and the truth of Einstein's relativity theory were difficult to accept. Each concept was, in its own right, a "stupendous intellectual voyage" which most people were unable to take. And there was a point at which even Einstein balked. To accept that the cosmos could be expanding was just too much for him.

Interesting. The assumptions, logic and mathematics that Einstein utilized to discover that the universe is expanding are accurately described as "airtight." Yet, Einstein found an expanding cosmos, implied by his own work, to be "irritating." To give such an idea a second thought was "senseless." Clearly, even Einstein was susceptible to cognitive dissonance. Einstein, the pillar of logic, suddenly became illogical.

Let us reiterate our findings. If a person is faced with information which either conflicts with popular views he had accepted, indicates he has erred or is complicated and difficult to grasp, the information represents an attack on his ego. As a result, the information "irritates." Psychologists call this irritation cognitive dissonance. Even though the information can be "scientific" and factually true, and the intellect should accept it and use it, the information creates cognitive dissonance, a purely emotional reaction. Because of cognitive dissonance, information which is factually and intellectually true can actually make a person squirm physically. As we have seen, such information is likely to be rejected out of hand, even if the one rejecting the truth is extremely intelligent.

BLACKOUT

The types of information discussed thus far—an advertisement for a product one chooses not to buy, Einstein's theory of relativity and scientific evidence about the shape of the earth or the size of the universe—were basically factual or intellectual types of information which had an indirect impact on emotions. Sometimes, however, new information can conflict with emotions directly. In such instances, the resulting cognitive dissonance can be so strong that it can totally prevent the information from even being heard! Sometimes, the information can be so emotionally upsetting that it is blocked out completely by cognitive dissonance; such information causes no physical irritation at all (no squirm).

When one finds oneself confronted by new information conflicting with one's emotional desires, a subconscious "early warning device" immediately shuttles the threatening information to a dusty corner of the mind, never to be heard from again. The mind "snaps," and suddenly, the threatening information disappears. Because of cognitive dissonance, emotionally disturbing facts can be entirely blocked out from a person's consciousness.

People can react to emotionally unsettling developments either by accepting them, and subsequently readjusting, or by rejecting them completely. Thus, if the implications of certain facts render life too much of a strain, our familiar blocking device switches on. The result is that the disturbing facts, despite their truth, never come to be evaluated.

Psychiatrist Rollo May writes in his best selling book, *Love and Will*:

> A patient of mine presented data, during the very first session, that his mother tried to abort him before he was born, that she then gave him over to an old-maid aunt to raise him for the first two years of his life, after which she left him in an orphanage, promising to visit him every Sunday but rarely putting in an appearance. Now, if I were to say to him—being naive enough to think it would do some good—"Your mother hated you," he would hear the words, but they might well have no meaning whatever for him. Sometimes, a vivid and impressive thing happens. Such a patient cannot even hear the word, such as "hate," even though the therapist repeats it. Suppose my patient is a psychologist or psychiatrist. He might then remark, "I realize all of this seems to say my mother didn't want me." He is not prevaricating or playing a game of hide-and-seek with me. It is simply a fact: the patient cannot permit himself to perceive the trauma until he is ready to take a stand toward it.

Certain facts carry implications which can simultaneously deliver both intellectual and emotional jolts. Usually, facts carry such potency only if people have been brainwashed against them.

To visualize this concept, we need only recall George Orwell's use of this phenomenon in his novel, *1984*. In the novel, a man being physically tortured by an official of "Big Brother" declares:

> "You people are after total domination, but you can't possibly succeed. You know why? Because the earth is only a speck of dust in the universe. So your whole life's aim is pointless."
>
> "What do you mean, the earth is only a speck of dust?" the Big Brother follower asked him.
>
> "Just walk outside," the answer came back, "and look up at the stars—millions, billions of them! And they're millions and billions of miles away. They're too far away for you. You'll never achieve total domination."
>
> "Oh no, you're mistaken," the Big Brother follower sadly replied. "These things that you call stars—they're just lights a few kilometers above the earth; we can shoot them down with our guns anytime we want."

Orwell's point was that Big Brother had succeeded in convincing his followers not only to adhere to a certain lifestyle but also to impose this lifestyle on others. Axiomatic to this lifestyle was the principle that Big Brother was the all-powerful, controlling force of the entire universe. Therefore, when the indoctrinated follower stated the obviously false maxim about the stars, he was not consciously fooling himself.

Even if he had wanted to, he could not have thought differently. He had no choice. He had been indoctrinated. Indeed, it would have been more than difficult for this follower to accept the truth about stars; it would have been impossible.

The idea that stars could be so far away and out of the reach of Big Brother was too contradictory and fundamentally threatening to his lifestyle and belief system; sweeping personal and social readjustments would have to take place if the idea were found to be true. Indoctrinated man subconsciously blots out the threatening idea, and thus saves himself the great pain of having to make major readjustments in almost every aspect of his life.

Once one has become indoctrinated to a certain belief system, adheres to it and lives by it, anything conflicting with that belief system will create strong cognitive dissonance. The conflicting idea will be perceived as a joke, and will quickly be laughed off, in a subconscious way, so as to preserve the advocate and his belief system, safely protected from the truth.

A person indoctrinated into an entire belief system has a huge "investment" to protect, so when faced with information threatening his investment, the cognitive dissonance is overwhelming. And often, the effects of the dissonance are bizarre.

However, as we have indicated from the start, one need not be indoctrinated into an entire belief system in order to experience dissonance and suffer its effects. Even one who has invested in a simple clock can experience dissonance. Dissonance can arise in any situation, as soon as there is an "investment," no matter how small, because man seems to have a subconscious need to "protect" his investments, even from the truth.

THE ANTI-TRUTH MISSILE

The upshot of all this is a bit unsettling. A person may be faced with an undeniable scientific observation, or some other fact or idea, yet because of this subconscious quirk in the human psyche, he may remain unconvinced! He might even deny the thing completely! To ward off facts, ideas or observations that are indisputably true but "hard to take," man has an extremely powerful subconscious defense mechanism, an "early warning system," which can intercept the incoming information and render it impotent. For emotional, sociological, philosophical, metaphysical and other reasons, information seen as threatening or "irritating" can be shot down and destroyed, or can be deflected into oblivion, keeping the information from ever getting to be evaluated by the human intellect. Even Einstein was susceptible to this human trait.

In other words, observations, facts and valid ideas are often rejected outright on grounds having nothing to do with logic or reason. The explanation for one's doubt might not be lack of evidence. Evidence sufficient to remove the doubt may exist, but the evidence might be "blocked out" by cognitive dissonance. And in the end, doubt remains.

CHAPTER TWO

COGNITIVE DISSONANCE AND G-D

When advocating the position that G-d created the world, one is often met not by outright denial but by an expression of non-commitment, such as the common reply, "I do not know."

But is such a response truly based on lack of facts?

There are two two types of "I do not know." One is the "I do not know" based on logic and reason. For example, before probes landed on Mars and sent back reports, if a scientist had been asked about life on Mars, he would have answered simply, "I do not know." The basis for this answer was purely rational. Scientists lacked information. Before the probes, scientists had no conclusive proof about life on Mars. There may have been life there, but no one could know. We will call this Type I.

Type II is the "I do not know" that is completely divorced from logic and reason. Doubt here is not based on a lack of evidence or information. On the contrary, the evidence here is compelling, but doubt springs from a very powerful subconscious "I can't take it." The mechanism creating doubt here can provoke bizarre, emotional and foolish behavior even among those noted for logic and reason.

EGO AND THE UNIVERSE

And so, in the following pages, the question will be: Is the "I do not know" concerning G-d a rational "I do not know," or is it a product of cognitive dissonance?

When a person refuses or is unable to accept G-d, is it because there is simply not enough evidence to prove that G-d exists, or is it because conclusive proof is available, but for certain reasons, the prospect is so wrenching, the person cannot accept it?

The question touches on religion, but its primary thrust is in the realm of human psychology: What goes on in the human mind when a person grapples with the concept of G-d?

Of course, one may say that believers in G-d lack good evidence to support their belief and simply *want* G-d to exist so much so that any evidence against G-d's existence produces dissonance in them and is disregarded.

Indeed, in today's secular world, most people look at the situation as so. They condescendingly halt discussion with those advocating the position that G-d created the world, (cutting short the discussion) with a sardonic remark such as, "Oh, well, if it's a matter of your religion, then I cannot argue." They regard the advocate of Creation as psychologically unprepared to follow through with pure reason.

On the other hand, by the same token, the converse may also be true. It is also possible that good evidence of G-d's existence is available, but people are non-believers because they are following what *they* want. Perhaps there is proof of G-d's existence, but because people have made precious emotional and intellectual "investments" in the position to the contrary, they succumb to dissonance and refuse to think logically.

Now, in order to entertain the suggestion that evidence of G-d's existence is, in fact, available but is "blocked out" by dissonance, one must first offer reason to suspect that evidence of G-d can create dissonance in the first place. That is, there must be a reason why people would be so irritated by the thought of G-d's existence that they would reject factual evidence supporting Him. Why should evidence in support of G-d have the power to create dissonance in the masses? Why should the idea of G-d make people uncomfortable? Why should it make people squirm?

Sometimes, the reason for feeling the irritation may be invalid, but as long as one merely *thinks* a reason for irritation exists, cognitive dissonance will result. This is apparent from several of the aforementioned cases, wherein cognitive dissonance was present, although it arose in the first place only because people were mistaken in some way and were lacking information.

For example, when people confronted by the new information about the shape of the earth experienced dissonance, it was only because they lacked key information about the force of gravity. As long as they did not understand gravity, they felt that if the new information was correct and the earth was, in fact, round, everything should be falling off. The information about the shape of the world was indisputably true, but due to a lack of information, dissonance arose and the new idea was rejected, even though the dissonance, in a sense, was "unjustified."

In other examples mentioned above, the dissonance that caused information to be blocked out and rejected was again ill-founded. It was born exclusively out of ignorance, out of shortcomings in human understanding. Take, for example, the dissonance which arose from the discoveries of Einstein or from the new telescopic evidence about the size of the universe.

When people were confronted by these new ideas, and actually squirmed physically and were irritated by them, it was error and lack of insight that spawned the cognitive dissonance. Nonetheless, the dissonance worked its effect, and the ideas were rejected outright, even though they were true.

In other instances, such as the case where a person was presented with evidence that his mother hated him, the dissonance produced by the information was "justified." Importantly, the evidence that his mother hated him was no better than the new evidence about the shape of the earth, nor any better than the new evidence about the size of the universe. That is, the evidence of the hate was not a truer reflection on reality than the evidence in the other cases. Even so, in the case of the mother's hate, the dissonance that the evidence produced was more "justified" and "well-founded," because in this instance, the dissonance did not owe its existence to any error or shortcoming in the realm of human understanding.

At any rate, whether or not dissonance in a particular instance is justified has no bearing whatsoever on whether or not the dissonance will perform its function in the human psyche and deflect and blot out what is seen as threatening. Even in the cases where the dissonance was unjustified and mistaken (because it arose exclusively from ignorance and lack of understanding), it was nevertheless present, and it discredited and destroyed the "incoming" truth as effectively and completely as "justified" dissonance.

Thus, when people are confronted by a new situation, idea or fact irritating them for some reason and making them squirm (producing dissonance in them), it does not matter, in a sense, whether they are right or wrong in their being irritated. The dissonance is present and works its effect just the same. Right or wrong, the irritation arising can block a person from all perception of truth.

Before any reasons are put forth for why evidence of G-d might create dissonance in people, it must be stressed that whether a person is "right" or "wrong" in being irritated by such evidence, the dissonance is very real, even if its cause is lack of information and insufficient understanding.

Right or wrong, why do people suffer cognitive dissonance when faced with evidence of G-d? Why might they be prevented from dealing with the evidence in a normal fashion?

Seemingly, cognitive dissonance can arise for five different reasons.

TROUBLE SPOTS

Evidence for the notion of G-d's existence can upset people emotionally because people tend to perceive the evidence as being in direct conflict with their personal and emotional desires.

Any evidence which supports the idea runs the risk of simply not being treated seriously; evidence of G-d might be blocked out of the consciousness completely, to prevent even the possibility of emotional stress.

1. People suspect that if G-d does, in fact, exist, then we as human beings might not be as free as we would like to be; G-d might have certain "do's" and "don'ts" for us, which we would be obligated to respect and follow. "Liberty or death" is but one saying which proclaims that man's drive for "freedom" is so strong it can override any other consideration. Because people are so attached to the idea of freedom, evidence of G-d can be irritating subconsciously, because the idea of G-d might be perceived as a threat to freedom.

A person might subconsciously tend to prefer the non-existence of G-d, out of fear for his own personal "sovereignty." Because people feel that G-d might want to interfere in their personal affairs, and seemingly limit their personal freedom, evidence of G-d can trigger considerable cognitive dissonance. Rather than admit that evidence of G-d might be valid, and as a result concede a certain loss of dearly held liberty, people subconsciously might blot out evidence of G-d so that they can remain sovereign, independent and unrestricted. In short, evidence of G-d can be emotionally upsetting, because it can make man feel small; the evidence implies that man may be limited in his personal freedom.

2. People also harbor a fear of discovering that they are but figments of a Creator's imagination, their existence precariously depending upon G-d's will. In the popular novel, *Breakfast of Champions*, Kurt Vonnegut depicts the kind of fear that is felt. At one point in the book, the author decides to descend into the

pages of his book to meet his favorite character, who is sitting at a bar, calmly nursing a drink. The character is suddenly overcome by tremendous anxiety and apprehension. He senses that something is about to enter the room, something "awesome" that he "cannot possibly face." It is the author.

Imagine the scene. There sits the character, perfectly content with the idea that he is a real human being. His encountering Vonnegut at that moment, would have grave implications toward his existential position, to say the least. Discovering that he is nothing more than a character in a story would require a tremendous adjustment. Can you appreciate the potential for trauma here?

Picture yourself as the character of that book. Instead of sitting at a bar, you and your friends are walking on a city street. Suddenly, it is revealed to you that no one in the entire world exists, not even yourself. You and your friends are nothing more than characters in a book, figments of some author's imagination. As a matter of fact, the scientists have just looked through the newest telescope and they have seen pages flipping. It is quite a horrifying thought.

Man is an expressive, creative force in the universe, and he takes great pride in this. Nothing shakes a human more than the idea that his entire being is, in reality, an invention of another expressive, creative force, a Being higher and more powerful than his own.

3. A person pondering the question of G-d's existence often will say to himself, "If G-d exists, and He is, in fact, a spiritual Father to all of us, why is G-d so aloof and obscure? Why doesn't He let us know He is here?" Rather than admit to evidence in support of the idea of G-d, and as a result suffer feelings of

abandonment, the human subconscious may blot out evidence of G-d's existence so that the person may be spared the emotional strain.

In other words, evidence of G-d's existence can create a sense of puniness and unimportance not only because the idea implies limitations on human freedom, or implies that man in relation to G-d is inferior, i.e., merely a creation. Evidence of G-d can also create a sense of puniness and unimportance because the idea can trigger a sense of abandonment and rejection. Just as people fear the thought of losing personal freedom, people also fear the thought of being rejected and abandoned.

For people who cannot understand why G-d should remain "behind the scenes," and who feel, rightly or wrongly, that as long as G-d is this way, He must be rejecting us, evidence of G-d can produce trauma, much as the trauma described by Rollo May in *Love and Will*. In fact, evidence of G-d might be blocked out completely in the subconscious, so that the dreaded feelings of abandonment never have the chance to arise.

4. If a person accepts G-d's existence, he must also admit to a lack of understanding. Thus, evidence of G-d can create dissonance in people for the same reason people experienced dissonance when faced with new information about the shape of the earth. Being that the earth's roundness did not "fit" easily with the hard fact that people do not fall off the earth, people accepted what was obvious to them—that no one falls off—and they rejected the new, abstract idea that the earth is round. If roundness had been accepted, people would not have been able to understand why they did not fall off, and the inability to understand would have belittled and upset them emotionally. As

people wondered, "If the earth really is round, why aren't we falling off?" people also wonder, "If G-d really exists, why are there so many problems in the world?" Rather than accept a new, abstract idea which seems to conflict with the obvious, and thus admit to a lack of understanding, people are prone to reject the idea subconsciously and save themselves embarrassment.

5. Finally, if one has "bought" the G-dless view of the universe in the marketplace of ideas and lived one's entire life accordingly, then clearly, evidence that G-d does exist will be difficult to accept, just as it is difficult to accept that one has bought incorrectly in the marketplace of products. The longer a person has lived according to the idea that G-d does not exist, the more dissonance there will be as a result of evidence to the contrary; the contrary evidence makes the person feel that much "smaller."

In sum, there are a number of reasons why evidence of G-d has the power to create great amounts of cognitive dissonance in people. Whether or not the dissonance is justified or based on a lack of information does not alter the reality that dissonance can arise and work its effects. Particularly because of its power to jar a person emotionally, evidence of G-d can be so disturbing that it might be blocked out completely, even in people who are highly intelligent. Because of dissonance, the evidence might be automatically dismissed in the subconscious, before the conscious intellect can rule upon it. Evidence in support of the idea of G-d might abound, but dissonance might be stopping people from seeing it.

Man's state of mind when he thinks about evidence of G-d might be analogous to the state of mind of a student sent to the zoo to write a term paper about lions. The student stands by the cage and watches. First, the student observes the lion's gait, his movements. The student takes notes. Next he studies the lion's interactions with other living things. The student takes more notes, and organizes all his thoughts, in complete calm. His concluding observations pertain to the behavior of the lion as feeding time approaches. Suddenly, the student notices that the door of the lion's cage is slightly ajar. At the same moment, he sees that the lion has noticed the door, too. The organization in the student's mind, his logic, his poise, are all immediately shattered. His whole being enters a state of emergency. No time for notes now! No time for orderly thinking at all!

This may be how it is with evidence of G-d. The implications of G-d's existence might be so unsettling subconsciously, that people might not get the chance to study the evidence calmly and logically. Evidence of G-d might trip the subconscious early warning device, and the evidence never penetrates to the mind. The evidence might be heard, but as Rollo May said, "it might well have no meaning" for the listener. For certain reasons, right or wrong, people might regard G-d as some sort of lion about to pounce. As a result, even though evidence of G-d's existence might be available, the intuitive and logical powers of the mind might be deprived of the opportunity to evaluate it.

SCIENCE'S ANNOYANCE WITH THE "BIG BANG"

Interestingly, cognitive dissonance over accepting G-d as Creator of the universe may actually be responsible for the initial rejection of yet another great scientific discovery. In this instance, scientists and laymen were not faced with new ideas about the shape of the earth or the size of the universe. Nor was it a question about the nature of light, energy or matter. On the table this time was a new theory about an even more fundamental subject—the origin and "beginning" of the entire universe.

For centuries, those who believed that the universe had a beginning were the "religious," the ones who believed in the account of creation found in the Bible. The rest felt that there was never a "beginning," and that what we see in the universe (e.g. the stars, the moon, the planets and how they move) always was as it is now, and probably always will be. This way of thinking came to be called "the steady state theory of the universe." Einstein believed in the Steady State Theory, which is why he tampered with his equations to keep the universe "static."

Those who adhered to the "steady state" idea asked a simple question: "Why assume that the universe was ever any different than it is today?" After all, astronomers had been observing the skies for thousands of years, and in all that time, there had been absolutely no change in the position of any star or constellation. The behavior of the earth, the moon and the planets had been "steady," too. Adherents to the Steady State Theory said that their way of thinking appealed to logic and reason, for it had been confirmed by scientific observation, while the idea of a "beginning" was essentially mythology, based on emotion, faith and religious dogmatism.

However, in 1946, a group of scientists led by George Gamow proposed a new theory agreeing in principle with the idea of a "beginning," and for the first time, the concept of "Creation" was said to be supportable by findings of hard science.

Basing their ideas on Einstein's theory of relativity and the astronomical observations of Hubble, Gamow's group proposed that at one time, not only were there no planets or stars, no earth and no moon, there were not even molecules or atoms! According to Gamow, the original "stuff" of the universe was a mysterious fireball made of pure energy, and the stars, the planets, the atoms and everything else came into being only after the original "fireball" exploded, in what Gamow called "the Big Bang."

Gamow arrived at his conclusion by using Einstein's famous formula, $E=mc^2$, which states that matter (m) can be converted into energy (E), which is the concept underlying the atom bomb. Gamow reasoned that the reverse should be true, too! That is, it should be possible to convert energy into matter! Just as matter can produce great amounts of energy (the atom bomb), a great amount of energy can produce great amounts of matter. Thus was born the idea of a mysterious "fireball" of pure energy having exploded, producing all the matter ever to exist in the entire universe. The idea seemed wild to many, but it successfully explained the observations by Hubble concerning the expanding universe. According to Gamow, the galaxies of the universe are rushing away from one another at high speeds in the aftermath of the Big Bang.

In 1948, colleagues of Gamow proposed that the Big Bang probably left an "echo," and they predicted that this echo might some day be detected although, according to the theory, the Big Bang occurred some 15 to 20 billion years ago.

As of 1948, there were still no instruments available that were sensitive enough to detect such an echo. But over the years, technology improved, and in 1965, two U.S. scientists were testing an extremely sensitive antenna meant to measure galactic radio waves. Unexpectedly, they detected weak electro-magnetic radiation that was coming, it seemed, from every direction in outer space. Although the two scientists, Arno Penzias and Robert Wilson, were not attempting to confirm Gamow's theory, they realized that the ubiquitous radiation they had discovered might be the "echo" Gamow's colleagues had hypothesized about in 1948. Penzias and Wilson announced their findings, and their measurements were subsequently confirmed by many other scientists. As a result, some backers of the Steady State Theory agreed that the faint background radiation in the universe is, in fact, the "echo" of the Big Bang. Subsequently, other key predictions of the Big Bang Theory were confirmed, such as predictions about the chemical composition of the universe.

Recognition of the Big Bang Theory got a special boost in 1978, when Penzias and Wilson were awarded the Nobel Prize in Physics for discovering the echo. (Nobel prizes are awarded only to the living, so George Gamow, having died in 1968, could not share in the honor.) By then, many backers of the Steady State Theory had conceded that the universe did have a beginning—a violent, explosive and mysterious one. The idea that the universe had a beginning was no longer said to belong only to the Bible. It was no longer dismissed as myth. With the awarding of the Nobel Prize to Penzias and Wilson, the idea became acknowledged as scientific fact.

Yet, as we mentioned, when Gamow's theory was first proposed, it was met not by mere reserve, but by great opposition in the scientific community. Even thirty years later, when Penzias and Wilson received the Nobel Prize, scientists were still bothered by the idea, for reasons which seem to relate directly to the idea of G-d. For one thing, according to Gamow, the mysterious fireball of energy existed before any matter did. If that pure energy did not emanate from G-d, from where did it originate? Moreover, if the entire universe at one point was simply energy, and there was no matter (not even protons, neutrons or electrons, and certainly no atoms or molecules), then these fundamental phenomena and all the many complex systems and processes observed today had to have developed, out of absolute chaos, in the finite period of time since the initial explosion. Given the magnitude of complexity on earth today, on the macro and micro levels, and given the interdependency between the many complex systems, it is difficult to understand how such development could take place.

The Steady State Theory posited that the universe always existed and always had a certain "inherent" order to it. Thus, it had been more conceivable that the complex systems and processes had developed on both the macro and the micro levels, and the processes of development would be unlimited.

In contrast, the Big Bang Theory calls for all order to have come from absolute chaos in limited time, and it calls for preservation of "acquired" order every step of the way, despite the natural tendency of order to decay. Such quick development of great order out of chaos is improbable, even impossible, without bringing G-d into the picture. Gamow's theory suggested G-d, and this is probably why the theory initially was very unpopular in the halls of science, and remained so for years.

Indeed, scientists, even more than laymen, were averse to accepting G-d's existence. Ever since the days of the Enlightenment, scientists have generally looked at their profession as a flight away from G-d. Science focuses exclusively on the physical, and its only route to knowledge is direct observation. For this reason, when Gamow, a fellow scientist, proposed a theory tending to support the idea of G-d, scientists fumed.

Not only did the Big Bang Theory support the idea of G-d's existence and creation of the universe, it even resembled the chronology of Creation given in the Bible. Just as it is stated in *Genesis* that G-d created light before He created the sun and the stars, which seems enigmatic, the Big Bang Theory proposes that the sun and the stars came into being only long *after* the initial explosion. They are merely "latter-day" by-products of the original pure ball of light and energy which appeared out of nothing. For this reason, too, the Big Bang Theory produced great amounts of cognitive dissonance, which blocked the scientists from accepting the theory for quite some time.

Discussing the Big Bang Theory, and why scientists were irritated by it, Robert Jastrow, director of NASA's Goddard Institute for Space Studies, had this to say:

> I think that part of the answer is that scientists cannot bear the thought of a natural phenomenon that cannot be explained, even with unlimited time and money. There is a kind of religion in science ... This religious faith of the scientist is violated by the discovery that the world had a beginning under the conditions in which the known laws of physics are not valid, and as a product of forces or circumstances we cannot discover. When that happens, the scientist has lost control. If he really examined the implications, he would be traumatized. As usual, when faced with trauma, the mind reacts by ignoring the implications.[11]

As noted above, even a layman experiences cognitive dissonance when faced with things he does not understand. A scientist, says Jastrow, especially wants to understand things and be able to explain them, so when faced with the Big Bang, something beyond his understanding, it is almost impossible for him to cope.

The main problem, though, was that the Big Bang Theory pointed to G-d, tending to confirm the account of Creation in the Bible. According to Jastrow, quoted in *Time* magazine:[11]

> For the scientist who has lived by his faith in the power of reason, the story ends like a bad dream. He has scaled the mountains of ignorance; he is about to conquer the highest peak; as he pulls himself over the final rock, he is greeted by theologians who have been sitting there for centuries.

In sum, one may be a shoemaker, a scientist, a vacuum cleaner salesman or a member of any other profession, and for any number of reasons experience dissonance when faced with evidence pointing to G-d. Whether the dissonance is justified is not the issue. For scientists in particular, there are yet additional reasons why evidence of G-d can produce dissonance. The scientist has invested in a belief system which he always regarded as being opposed to the idea of G-d, and to discover that he was mistaken is very unsettling to him.

THE ANTHROPIC PRINCIPLE

In the 1980s, there arose a movement, a sort of "school of thought" composed of professional scientists who believe in G-d precisely because of science. This movement espouses what is called the "anthropic principle," which suggests that the universe is "man-centered," brought into being by G-d for the sole purpose of generating human intelligence. According to this principle, G-d desires an intelligence other than His to exist, an intelligence capable of appreciating the universe in all its amazing design.

Naturally, the emergence of this new school of thought has literally incensed the professional scientists who do not advocate it. Critics of the new school maintain that the question of how the universe came into being is a scientific question, and to say that G-d is responsible is not "scientific," because the hypothesis cannot be tested. Furthermore, the critics say, to assume that G-d did create the universe and then to speculate as to why, is even further beyond the bounds of science.

Many of the scientists who have swung over to G-d have written entire books providing the scientific justification for why they have done so. The writings are voluminous and usually of a technical nature, making a comprehensive presentation beyond the scope of this work. Later, however, the general nature of the justification will be discussed, and we will illustrate it by means of a few examples.

This new development—that many professional scientists have come to believe in G-d—is significant, because the swing to G-d has been made solely on the basis of scientific findings. When the believing scientists are criticized by their professional colleagues who do not believe in G-d, the criticism may arise more from cognitive dissonance than from intellect and reason.

For now, however, let us get back to people who are not

professional scientists. Let us get back to the shoemaker, the vacuum cleaner salesman, and the people in other walks of life who are not privy to the latest scientific discoveries and do not have the biases and special interests which are particular to scientists. Let us take average people, men and women. What goes on in the human being's mind when it grapples with the issue of G-d's existence?

CHAPTER THREE

THE OBVIOUS PROOF

I n the modern era, since the time of the Enlightenment, people from all walks of life have spent entire lifetimes puzzling over the issue of G-d. Why have people been perplexed about whether or not the universe is a creation of G-d,? When people say, "I do not know" about the question of G-d, is the "I do not know" a Type I or a Type II? Is it based on logic and reason, on lack of evidence, or is the evidence of G-d sufficient, and the doubt a product of cognitive dissonance?

The best-known approach of those who support the existence of G-d is the classical "clock in the desert" argument, known also as the "argument for design." The argument says, "If you were walking alone in a desert, a place where, you were told, no one else had ever stepped foot, and you suddenly came across a clock, would you still believe you were the first intelligent being to have been there? Would you not conclude that someone had been there before?"

Why?

Because it is manifestly absurd to imagine that all the intricate parts of the clock, the gears, springs, hands, numbers and casing, could have just happened to fall into place, carried by winds and put together to form the perfectly harmonized mechanism. It obviously was manufactured by someone who knew about telling time and about engineering an instrument to suit the purpose.

In other words, the design of something proves it had an intelligent designer. Anyone seriously suggesting that a manifestly designed object had just "evolved" would justifiably be considered less than sane.

Let us restate the argument with an example closer to our daily lives. In the neighborhood grocery store, all like products are stacked neatly together on their own shelves. All the different brands of canned tomatoes are on one shelf, all the canned corn are on another, the canned fruits are on a third shelf, the baked goods are on a fourth shelf, dairy products are on a fifth, and so on. Certainly, everyone agrees that such organization cannot come about by chance, a lucky wind. Everyone knows that such order and design can be created *only* by intelligence. In the world of nature, which exhibits even greater design than the shelves of a grocery store, does it not follow that there must also be a designer?

The argument is airtight. Yet somehow, people are not convinced by the "proof" from the design theory. They walk away feeling that they lack the required information for an intelligent decision.

The design theory is one of the many arguments supporting the existence of G-d. The prevailing opinion, however, is that the design theory and the others do not provide sufficient data. Thus, the "I don't know" regarding G-d's existence generally is thought to be a Type I "I don't know." On the other hand, perhaps the doubt arises not from logic and reason but from a subconscious and irrational "I can't take it." Perhaps it is a Type II doubt.

Does the classical "clock in the desert" argument for G-d's existence fail to offer sufficient evidence to bring man an intuitive appreciation that an Intelligence created the universe? If we can demonstrate that it does not fail to do so, if we can show it is logically compelling, we may conclude that a person's failure to believe in G-d stems only from cognitive dissonance.

DESIGN'S "THRESHOLD"

First, we must establish the level of design required to prompt the average man to react automatically that "this is a product of intelligence." In other words, we'll have to perform an experiment that establishes the level of structured complexity which brings the average man to conclude intuitively, "This object did not come into existence by chance." We shall call this level of complexity the "threshold for design."

To discover the threshold, we will have to set up a situation free of the factors able to cause a triggering of the subconscious "early warning device." We will need an experimental setting where some level of design is present, and our subjects are under no personal, social, intellectual, metaphysical or other pressures at all which would prevent their seeing it. In other words, we will need a controlled environment, that is, a situation lacking any factors liable to interfere with the normal functioning of man's intuitive faculty. By means of a graph, we can illustrate clearly what we mean by the term, "threshold for design."

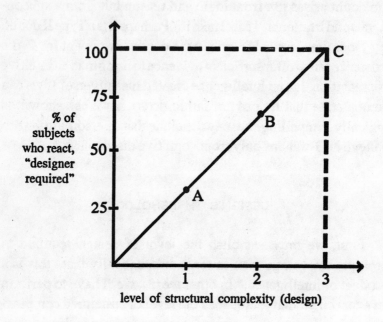

In this graph, the horizontal axis pictures increasing levels of structural complexity. The vertical axis indicates, in terms of percentage, the number of persons in a hypothetical group who respond that the levels on the horizontal axis could not have come about by chance. Here, only Level 3 would qualify as a "threshold for design." Level 3 represents the structural complexity where 100% of our subjects agree that "this could not have happened by chance." At level 2, less than 75% of our subjects react this way, and at Level 1, the percentage is even smaller.

This graph might represent the results of a controlled experiment involving university students and playing cards. Let us say Levels 1, 2 and 3 correspond to three different occasions, or "runs," wherein one thousand students are given playing cards. After each "run," students are asked individually, "Were the cards you received shuffled randomly before you got them, or had someone arranged them in a certain order?" On the first run, let us assume, each student received fifty-two cards, and all the hearts and spades appeared in ascending order from Ace to King. Given this level of complexity (Level 1), about 35% of the students (Point A) concluded, "Somebody arranged these cards."

On the second run, let us assume, each student again received a complete deck of fifty-two cards, but upon examination, each found that the hearts, spades, diamonds and clubs were grouped separately, and in each group, the thirteen cards appeared in ascending order from Ace to King. Given this level of complexity (Level 2), almost 75% of the subjects concluded, "Somebody arranged these cards . . . there is a designer here."

On the third run, let us assume, students found the same arrangement, but this time there were two decks, not one. Given this level of complexity (Level 3), every single student agreed, "Designer required."

Two objections are frequently heard regarding experiments of this sort. First, not enough subjects were tested. Second, the subjects have peculiar characteristics, and therefore they are not representative of all people. In other words, results from the experiment involving the students and the cards might be challenged because only one thousand students were tested, and because students are a special, non-characteristic group.

Fortunately, a quality experiment establishing the level of

complexity needed to provoke the intuitive reaction "designer required" already has been conducted. The controlled environment was the everyday cinema, and the subjects of the experiment were the millions of people who saw the classic film called, *2001, A Space Odyssey.*

THE *2001* MONOLITH

The film begins with highly artistic suggestions that man evolved from monkeys. A troupe of apes is living by a water hole, assumedly on earth. The *2001* director and make-up man having done a fantastic job, the movie audience is treated to a close-up view of ape society. Different "interpersonal" relationships are portrayed. It is even possible to recognize differences in character between one ape and another.

Just as the audience becomes acquainted with the first troupe of apes, a second troupe attacks and takes over the water hole. The audience follows the first troupe into exile, where at night, the apes find a cave for shelter. After a fearful, sleepless night, the apes find a strange object, totally out of place in ape society. It is a fifteen-foot high monolith, a perfectly rectangular black slab, shaped like a domino or the United Nations building, with perfect right angles and smooth, polished surfaces. After working up enough courage to draw near it, and after stroking the smooth surfaces with fear and wonder, the apes move on.

Afterwards, one of the apes makes an important discovery. While playing idly with a large bone from the skeleton of a dead animal, the ape realizes that the bone can be used to break other bones. He arms his comrades with the new weaponry, and the first troupe retakes the water hole, routing the attackers. There is no explanation regarding the monolith, and suddenly, the scene switches to spaceships, an example of the sophisticated technology of the twenty-first century. The United States has a colony on the moon. The juxtaposition of scenes implies that civilization on earth has made great strides forward in the many years since the first technological advance of discovering the military value of bones.

Suddenly, the colony breaks communications with earth. The colony, rumors circulate, has been hit by a disease of epidemic proportions. But these rumors are actually a cover-up. The real reason for the break in communications is a discovery on the moon with "such implications for social shock, the average man will not be able to take it." Better not to let earth know. For, *the Americans have found the first objective proof that the galaxy contains intelligent life other than man.*

What is the proof? Digging beneath the moon's surface, the human explorers have found the exact type of monolith the apes had found. Earthmen had no part in its being placed there, and it is therefore "the first objective evidence of any intelligence in the universe other than man."

First, note that not one character in the film is portrayed as objecting to the conclusion, based upon finding the rectangular slab on the moon, that "this is . . . objective evidence of . . . intelligence in the universe other than man." Nor has any reviewer or critic objected to the story's logic, criticizing the underlying assumption of the story as unreasonable. Has *anyone*, during or after viewing the film, objected to the logic behind the conclusion? No. There, in the theater, eating popcorn, free of personal, social, intellectual or other biases, people agree unanimously that a black slab with right angles and smooth surfaces is conclusive proof of an intelligent designer. In the theater, a controlled environment, there was no interference. Cognitive dissonance was absent. Not one viewer maintained, "Maybe the slab just happened." Every viewer had the same immediate, intuitive, "gut" reaction that the slab was proof of an intelligent originator. There was no doubt whatsoever.

Now, 2001 was viewed by millions of people from all walks of life, so it cannot be argued that too few people were "tested" or that the subjects of the "experiment" were not representative. Therefore, what level of complexity does it take for people to see intuitively that something was made purposefully? Does it take a computer found on the moon? No. A car? No. A *clock*? No! A simple black slab is enough. 2001 serves as a controlled, scientific experiment establishing man's intuitive "threshold for design." In the theater, where there are no implications for one's life, this threshold level is quite low.

This threshold level is usable as a reference point for learning the true nature of man's doubt about G-d. We can compare it to the level of design which is manifest in the universe. If the design in the universe is inferior to the design of the slab, if the level of design in the universe is below the threshold, we will be forced to conclude that the design argument offers insufficient evidence that the universe was designed. Modern man's "I do not know" regarding G-d would then be classifiable as a Type I expression of uncertainty. The doubt would be based on logic and reason.

On the other hand, if the design in the universe is superior to the design of the slab, if the level of design in the universe is higher than the threshold, we will be forced to conclude that sufficient evidence of a Master Designer is available. And, were it not for personal, social and other bias, in a word, dissonance, people would realize intuitively that a Master Designer exists. In this eventuality, man's "I do not know" regarding G-d would have to be classified as a Type II response. The doubt would be based on the subconscious, irrational "I cannot take it."

CONCEIVING OF CONCEPTION

Those with even casual knowledge of biology or the other natural sciences are well aware that nature's level of complexity far surpasses the complexity exhibited by an object with right angles and smooth surfaces. Let us appreciate this. An embryology textbook, *From Conception to Birth*,[12] asks honest questions about the human brain and its nervous system. How do the billions of cells comprising this system come into being in the first place, the textbook asks, and how do they attach themselves to each other to form a network connecting the brain to every muscle, organ and gland in the entire body? How is it possible for microscopic chromosomes, each containing all the coded information necessary to produce and "wire" an entire human being, to have come into existence without a designer?

It would be relatively easy to understand if the neurons were connected to the brain like spokes of a wheel, but they are not. Most of these neurons are connected to a great many other neurons; one estimate is that, on the average, each neuron is cross-connected with one thousand others. This means a total of ten trillion connections. A complete wiring diagram of this network would stagger the imagination. All of the telephone cables of the world would comprise no more than a small fraction of it.

The neuron, like any other cell, contains a nucleus in which lie chromosomes that are identical to those in the original fertilized egg. Thus, the nucleus of each neuron contains a catalogue of potentiality inherited from both mother and father. How can a collection of genes possibly account for the multifarious connections between neurons and the human nervous system? Or for the relationship between neurons and the muscles and organs of the body? There are only approximately forty thousand genes in all the chromosomes, seemingly not enough to encode instructions for performing ten trillion connections.

But if every last interconnection is not spelled out in the chromosomes, then how *do* the neurons get connected? Do they just reach out for one another haphazardly? Obviously not, since all neurons fulfill definite, specialized functions, not random ones. Connections between the nerves associated with hearing and those controlling, say, the bicep muscles wouldn't be logical or effective, and above all the nervous system effectively coordinates whatever the person does or thinks . . .

The nervous system eventually comprises the most efficient cable system in the world for the transmission of messages. Ultimately, each nerve fiber will be covered by a sheath of protective cells (sometimes five thousand per fiber), and each will be able to carry messages at a speed of one hundred fifty yards per second, or three hundred miles per hour. From these primitive cells, first distinguishable at eighteen days after conception, the embryo will form more than ten thousand taste buds in its mouth . . .

Some twelve million nerve endings will form in the baby's nose to help it to detect fragrances or odors in the air. More than one hundred thousand nerve cells will be devoted to reacting to Beethoven's Fifth Symphony or the ticking of a Swiss watch. The piano has only two hundred and forty strings, but the baby's ears will have over two hundred and forty thousand hearing units to detect the smallest variations in sound.

The baby's eyes, which begin to form at nineteen days, will have more than twelve million screen points per square centimeter; the retina, or light-sensitive portion of its eye, will have more than fifty billion such points. The composite picture the eyes record is homogeneous, because these light-sensitive points blend into a whole. Take a hand lens and examine any picture in any daily newspaper. You will find it made up of hundreds of points, each light or dark, which together make up the picture as you look at it from a greater distance. This is exactly what the eye does, only in much finer detail.

Where do these billions of cells in the nervous system come from? From the original fertilized ovum, which is still dividing after one month to form the tissues and organs that the child requires. It has been estimated that all two billion of the specific nerve cells which make any individual educable are located in the outer covering of his brain, its cortex, and that these two billion cells could be stored in a thimble.

Development continues in certain parts of the brain, even after birth. By the end of the first month of embryonic development, none of these parts of the brain, spinal cord, nerves or sense organs is completely formed, but the foundation for all of them has been laid.

The textbook does not mention the word "G-d," but it states:

The development of the brain and nervous system and its rule of the integration of all the systems remain one of the most profound mysteries of embryology.

The eyes alone display such intelligent planning as to stupefy anyone studying them:

The eyes . . . for example, are formed on the sides of the head and are ready for connection to the optic nerves growing out independently from the brain. The forces that ensure this integration have thus far not been discovered, but they must be formidable indeed, since more than one million optic nerve fibers must mesh with each eye. Think for a moment about what is considered to be a feat of human engineering: the drilling of tunnels from both sides of the Alps that must somehow meet precisely and merge into one continuous highway. Yet any one of the thousands of things the fetus must do as part of the routine of development is far more wondrous.

The eye has two types of receptors—rods and cones, located on the retina, the eye's rear wall. Image data collected by the rods and cones is somehow converted to electrical energy. Afterwards, in the retina itself, several distinct layers of tissue must be crossed before the data is conveyed to the brain via the optic nerve.

At first, vision theory held that each receptor's data would pass from layer to layer by means of its own distinct line of nerves, and once out of the retina, each receptor's data would go to the brain on its own distinct fiber in the optic nerve. Also, the focused eye was thought to be stationary, so that as a person looks at a tree, for example, only one small point on the tree would be seen by any receptor. On account of the many receptors, however, each with its own "line" to the brain, the brain was said to receive enough quality data to be able to recreate the image of the entire tree accurately, allowing the tree to be seen. According to all this, the job of the eye was simple. It was to collect the receptor data and send it to the brain. All "processing" of the data was said to be done in the brain, not in the eye.

However, this theory has been discredited on several counts. First, data from a receptor is almost never sent through the retina on a single line of nerves through the many layers. Second, the number of receptors on each retina, though more than fifty million, is not fifty billion, as originally thought and as quoted above. In comparison to the number of receptors in the two eyes (over one hundred million), the number of fibers or "lines" in the optic nerve (fifty thousand) is quite low. From this alone, it is obvious that data from the receptors is "processed" in the eye itself before it continues and reaches the optic nerve.[13]

Furthermore, it now is known that the focused eye is not stationary, as was believed. Rather, the focused eye moves in an orbital fashion, randomly, around its point of focus. Also, built onto this random movement is a regular movement, at the rate of 30 cycles per second, peaking every ten seconds. As a result of these two different movements, each receptor sees not only one point on the tree, but a whole section. Receptor cell number 10,001 sees one section of the tree, and cell number 10,002 sees another section; but what the second cell sees overlaps what the first cell sees.

To sort out the constantly changing and redundant data from the many receptors, the eye itself has processing centers in layers of the retina once considered to be void of significant structures. In these microscopic processing centers, the great mass of receptor data is made sensible, condensed and finally adapted to the limited number of "lines" in the optic nerve. In the brain, the coded message is reconstructed, so the eye is able to see.

Now, for all this to work, the eye needs enormous input from many outside sources. For example, it needs energy, nutrition, oxygen and lubrication. Also, the pressure and temperature inside the head must be within certain critical limits. Numerous factors must be just so. In short, were it not for proper functioning of thousands of other systems within the body, the eye would simply not function at all.

Within the eye itself, vision is not possible unless billions of component parts do their job properly. There is the lens, for example, the light-admitting pupil and its iris, the focusing mechanism and the retina. In the retina alone, there are millions of receptors, other cell types, complicated processing centers and numerous substances to help conduct the current from structure to structure.

Staggering numbers of interdependent components are found not only in the eye, concerned with vision, but in other parts of the human body, too. And as with the eye, the systems employed to accomplish all the vital tasks of life depend on proper operation of countless components. The chances of malfunction are incredibly high. It is a major miracle that a person wakes up in the morning.

No one will deny that the human body's complexity far surpasses the threshold level provided by the black slab in

2001. Therefore, the obvious question is as follows: If, when in the theater, a person's intuition tells him that a black slab with right angles and smooth surfaces had to have been purposefully made, why doesn't that person's intuition tell him the same thing when he looks at a newborn or at a potted plant?

Clearly, this is a prime example of the principle of cognitive dissonance. The existence of the obviously designed creatures of the universe imply an intelligent Creator, an uncomfortable proposition to those who have made large "investments" elsewhere. Those who doubt or deny G-d are missing the truth only because the truth of G-d's existence is subconsciously irritating to them.

In other words, doubt or denial of G-d is not based on logic, reason and lack of evidence. We are not really dealing with an "I do not know." We are dealing with an "I cannot take it." Through this camouflage, nature's overwhelming evidence of G-d is deflected into oblivion so that the investments are protected.

THE ULTIMATE PROGRAM

People do not usually appreciate the sophisticated level of design in their very own beings. Allow us to illustrate.

Imagine it is the year 2000, and American businessmen are so outraged by the proliferation of costly environmental controls, they begin a movement to reproduce American industry elsewhere, where environmental considerations will not be a problem, namely, on the moon.

Industry magnates approach the American space program, a

warehouse of know-how. Their request raises several eyebrows: "Make us a space capsule able to land softly on the moon, without anybody on board. The capsule need only get there. It need not come back. However, we would like it to have a computer equipped with a special program containing instructions for acquiring all the raw materials necessary to assemble tools to recreate American industry on the moon. In the program, there must be a 'stop' once the raw materials are collected.

"For the next step, the program must contain instructions for building the tools and the machines of industry. Then there must be another 'stop.' Next, the factories will be assembled, as well as supply lines and power lines between them. Then there will be another 'stop.' Next, the program must ensure that the entire creation has internal surveillance systems for any malfunction, as well as systems for maintenance and repair.

"After this step is accomplished, everything must be turned on and industry will spring into action. Once everything is operating smoothly, the computer must reproduce itself, its program, and also another space capsule, and all this must blast off to another planet to recreate American industry there."

Any computer programmer will state unhesitatingly that it is not even remotely possible to program a computer to accomplish this. The scheme is far beyond the wildest science fiction. However, embryologists and others agree, this is more or less the sort of program contained within each and every human being.

The first step in the human "program" is the uniting of the male sperm with the female egg. After the coupling, the fertilized egg nestles into the side of the uterus, and automatically, raw materials begin to be assembled. These materials are used to make proteins and other substances, that is, "tools."

Next, the program sees to it that the tools build cells, that is, "machines." The machines are then put to work assembling organs, that is, "factories." One factory produces insulin, another makes lymph. Transportation and disposal systems are established, as well as systems to detect infections and malfunctions. Other systems are dedicated to repair and maintenance.

After billions of ordered steps, the factories (organs) are able to work independently, but really, each one depends upon the services and products of other "factories." Inside man, therefore, is a computer-like program assembling high level "technology" from scratch. Included in the program is foresight for the future, for a time when the new technology will be mature and running as planned. That is, a baby eventually will be able to recreate itself. In the program there is provision for exactly how human "machinery" can appear again in another location!

Nature works with its materials, and man works with his, but when the results are set before us, man must blush. Design in nature clearly exists.

EMPIRICAL PROOF

Thus, audience reaction to the film *2001* belies those who claim G-d is "an invention of man," an imagined concoction that man needs in order to help himself cope with life's ubiquitous difficulties. The truth of the matter is quite the opposite.

If the question is approached with full objectivity, it becomes perfectly clear that nature provides overwhelming evidence of G-d! Design in live objects in nature is much more sophisticated than the design of the *2001* slab, and as people recognize intuitively, even the slab could not come about without the intervention of intelligence. In the movie theater, where there are no implications for one's life, people recognize the truth easily, because there is no emotional dissonance to stop them from seeing it. It is, therefore, clear that those who do not believe in G-d are the ones who are suffering cognitive dissonance.

FLASHBACK

Now let us backtrack. The argument goes as follows: In light of the very high level of complexity in living things, people should infer that living things were made by intelligence. After all, non-living things such as rectangular slabs and clocks are not nearly as complex as anything alive, and only intelligence can produce rectangular slabs and clocks.

If the complexity in non-living objects does not come about without the intervention of intelligence, does it make sense to say that the greater complexity in live objects does? No! People would acknowledge intuitively that intelligence created live objects, but deep-seated subconscious blocks stand in the way.

AGNOSTICISM ON THE DEFENSIVE

Skeptics are apt to claim that the argument for design has long been refuted by philosophers and intellectuals. However, this is simply untrue. We will demonstrate that the proof has never been refuted in the least.

A philosophical objection to the "clock in the desert" argument" asserts that human beings are unable to perceive or recognize anything, including design, without the benefit of experience. And, the argument goes, since different people bring different experiences to any given situation, perception is totally subjective. According to this, the perception that anything is designed is nothing more than a personal opinion based on one's particular experiences. In stressing the importance of experience, as well as its subjectivity, this view completely denies that man has an intuitive power to recognize and appreciate reality as it actually is.

A related argument asserts that living forms are in many ways different from non-living forms, and any conclusion drawn about inanimate objects is not necessarily applicable to something that is alive. This philosophic argument goes as follows: When people look at clocks, tables or monoliths, they are able to conclude that "this was designed" only because, in the past, they have actually seen such things being made. The manufacturing process is part of human experience.

On the other hand, when people look at babies or potted plants, all they can recall from experience is that these things seem to take on design by themselves, spontaneously. A person may have a vague feeling that the design in babies and plants is, in truth, the work of a designer. However, because he lacks the experience of having seen these things actually being made, he is not sure. Therefore, according to this, a person's not recognizing a designer behind babies and plants does not necessarily stem from personal or social blocks which cause a malfunctioning in his normal perceptive powers. Rather, non-recognition stems from a lack of experience.

For these reasons, skeptics about Creation reject it as the explanation for design-filled life on earth and instead become faithful believers in evolution, as popularized by Charles Darwin. According to this, babies and plants came into being by chance, spontaneously, without the intervention of a Designer. Rather than attribute design in nature to Intelligence, subscribers to evolution attribute it to "natural selection."

CHAPTER FOUR

IS DESIGN A FIGMENT OF OUR IMAGINATION?

At this point let us focus on the two objections we have just listed. Raising the first objection, some philosophers would sabotage our entire line of reasoning by suggesting that design may be only a dream. They claim that a person is incapable of judging whether seeds, photosynthesis and other immensely complicated qualities of the tree or of the embryo, are really created by an intelligence. Man, they assert, has no basis for contending that the world exhibits compelling evidence of design.

They also postulate the existence of other universes, of which Man is totally ignorant, which may very well function without what we perceive as design. Even our universe may be designless, never having been the product of a designer. One's perception of design cannot be trusted, and therefore, there is no basis for a proof of design in the universe.

The second of the above objections claims that humans

ascribe purpose and plan to the world out of a psychological need. The idea of design brings order and sanity to our lives. The conclusion that the world was designed stems from a strong need to believe it to be so, and not from objective and compelling reasoning.

According to this school of thought, man is small, weak, very limited, not able to judge and certainly not able to generalize with any degree of dependability. For the sake of convenience, let us call this the "Myopic Man" school of thought. It claims that man cannot be trusted to say there is design in the universe as a whole or in natural objects viewed individually.

However, scientists do not consider design a dream. Ever since 1978, when Penzias and Wilson were awarded the Nobel Prize in physics for discovering what is considered by many to be the "echo" of the Big Bang, the move of scientists to G-d has picked up steam, because scientists are finding that the universe "as a whole" is such an incredible combination of interacting factors, it is impossible for it to have developed from the original "fireball," were it not for guidance from "outside."

The atoms, stars, galaxies and living beings had to have stemmed from the original "bang," but the development of each phenomenon depends on a wide range of unlikely coincidences. Many scientists agree it had to have been guided, and it is difficult to imagine where that "outside" guidance came from if not from G-d. One scientist writes:[14]

As we look out into the universe and identify the many accidents of physics and astronomy that have worked together to our benefit, it almost seems as if the universe must, in some sense, have known that we were coming.

According to Paul Davies,[15] a professor of theoretical physics:

It is difficult not to be struck by some of the surprisingly fortuitous accidents without which our existence would not be possible.

Some of the happy "accidents" involve numerical "constants," which describe unchanging fundamental "laws" of nature. Davies writes:

As more and more physical systems, from nuclei to galaxies, have become better understood, scientists have begun to realize that many characteristics of these systems are remarkably sensitive to the precise values of the fundamental constants. Had nature opted for a slightly different set of numbers (i.e., if even one of the constants were "off" by even a small percent) the world would be a very different place. Probably we would not be here to see it.

More intriguing still, certain structures, such as solar-type stars (our sun), depend for their characteristic features on wildly improbable numerical accidents that combine together fundamental constants from distinct branches of physics (the existence of the sun depends upon not one "accident," but several).

And when one goes on to a study of cosmology—the overall structure and evolution of the entire universe—incredulity mounts.

Recent discoveries about the primeval cosmos (the universe from the time of the Big Bang) oblige us to accept the fact that the expanding universe has been set up in its motion with a cooperation of astonishing precision.

A description of all the different "coincidences" is beyond the scope of this work. The fact is, the argument for G-d which bases itself on the natural "constants" is not dependent upon the Big Bang Theory. Even if the Big Bang Theory is some day refuted, still, for many scientists, the natural "constants" by themselves are proof that the universe is the work of G-d.

Presently, the Big Bang Theory is, in fact, widely accepted, and together with what is known about the natural "constants," scientists are saying that man can, in fact, conclude "designer" on the basis of the universe "as a whole." This means that one of the major contentions of the Myopic Man school, the distrust of man's perception of design in the universe "as a whole," is flatly denied in today's halls of science.

The findings of modern science also cast grave doubts on the Myopic Man school's idea of "other universes." If all the atoms, stars, galaxies and other phenomena of the universe did originate in the Big Bang, who is to say there ever was a big bang elsewhere? Where "else" could such a second big bang have taken place?

Moreover, if atoms, stars and everything else in our universe originated in our Big Bang and congealed into sensible structures only because of an incredible string of coincidences, then even if there had been some other big bang "someplace else" outside our universe, who is to say that such a hypothetical big bang would produce anything in the absence of the needed coincidences? It is difficult to imagine what another universe would be made of, because without intelligent guidance, it is highly unlikely that such a proposed "other" big bang would produce anything but chaos.

In sum, modern science indicates that man can conclude that there is a designer on the basis of viewing the universe "as a whole," and the conclusion should not be doubted on the grounds of possible "other universes," because it is extremely doubtful that such "other universes" could form or even have a place to form themselves. Seemingly, when critics of the designer concept propose mysterious, unseen universes other than our own, they are creating a mythology because of cognitive dissonance, in order to escape from the simple truth.

All this notwithstanding, let us turn now to the other major contention of the Myopic Man school of thought, the contention that man cannot conclude that a designer is behind natural objects viewed *individually*. We will examine the characteristics of distinct, separate natural objects within our environment in stark isolation and show that the design in such objects is an objective reality.

The objects we will show "speak" to us, saying, "I am a product of purpose and plan." We cannot help but conclude that man would be able to hear this clearly, were it not for certain intellectual blocks, as well as man's subconscious "early warning device." In other words, it is possible to demonstrate that people are not "reading design *into*" things. Design in nature is clearly real, and man's perception of the design can be trusted in full.

THE IMAGINARY SPACE SHUTTLE

Everyone must agree that a brand new automobile exhibits a certain level of structured complexity and that this complexity did

73

not arise spontaneously. The auto was designed. It was given hundreds of working parts, each with a specific function. Some parts work to supply the engine with fuel or with sparks. Other parts help with lubrication or with cooling. Then there are parts for steering or for stopping. When all these parts function together properly, the car not only starts, but it runs. Smooth running was what the designers had in mind.

Now, if ten thousand such cars were rolled out onto a huge parking lot, and ten thousand drivers got behind the wheels of the cars, despite the fact that every car would be brand new, it is quite likely at least two or three of these cars would not start, or they would quickly break down.

This is so precisely because each car is composed of so many interdependent parts. If one part fails to perform exactly as it should, it often means that others will malfunction, too. In fact, a problem with a single part can cripple the whole car, keeping it from starting or running.

Must the car be built this way?

The answer is yes.

The car is built with many interdependent parts because this arrangement is required for the car to accomplish all the things it is supposed to do. Now, as compared to an auto from model year 1960, the new model is built to accomplish much more. It has a pollution control system, for example, or automatic speed control. Therefore, the new car has more parts than the one built in the year 1960.

Surely, though, designers aim for simplicity. If they could build a new car with as few parts as a 1960 model, yet not sacrifice the added features, they would. Nevertheless, whether we're speaking of the newer or the older models, if ten thousand automobiles were rolled onto a parking lot, it is the nature of machinery that at least a few of these automobiles would not start or run.

As noted, a car has parts which supply fuel to the engine, others which supply sparks, and others which help with lubrication, cooling, etc. These parts are essential if the automobile is to do what it is supposed to do. If a new automobile performs more tasks than an older 1960 model, and therefore requires more parts, then obviously a spaceship, which must perform so many more complex tasks, requires many more parts than any ordinary automobile.

In fact, a legitimate analogy would be to liken a spaceship to ten thousand automobiles in a parking lot, all connected together into one machine. The task of the spaceship is so difficult to accomplish that ten thousand interdependent systems are absolutely essential to the spaceship's design. As with an automobile, a problem with a single part can cripple the entire sophisticated machine. In the spaceship, however, since there are so many more complex and varied parts, the chances for malfunction are that much greater.

Furthermore, a spaceship has a "self-surveillance" system that is able to detect future malfunctions and, on that basis, halt a countdown. If, when ten thousand cars are produced, people expect that some will not start or run, then certainly, when a spaceship is standing on the launch pad, with its surveillance systems on guard, observers should expect some occasional "holds" in the countdown.

The high level of design required for spaceships should be something Americans, of all people, should appreciate. Often, when the United States space program was attempting to launch its space shuttle, the countdown was repeatedly halted. Every time the team attempted to have the shuttle blast off, some part somewhere inside the ship would malfunction, or threaten to malfunction. While a disappointed American public sighed and returned to its affairs, the troublesome part would be repaired or replaced. This happened again and again.

However, if the American public had appreciated the analogy to ten thousand interconnected cars standing in a parking lot, where if one fails to start, none start, the Americans would have been much more understanding.

The point is, as the tasks given to machines get progressively more difficult (the task of the shuttle was much more difficult than the task of an auto), the level of required design rises proportionately, and so do the chances for malfunction. Greater challenges demand greater complexity, and greater complexity increases the likelihood of failure. Therefore, if a person pauses to reflect, it was a minor miracle the space shuttle lifted off at all.

This proves that the design in autos and spaceships cannot be imaginary. Could Lithuania build a space shuttle? Could Morocco? In 1960, could America? What we are saying is really quite simple: Although "design" is itself an abstract concept, there are levels of design in the real world. Design has hierarchies. Some tasks call for levels of design which are achievable only by a select few technological giants of the modern age.

In other words, the Lithuania and the Morocco of today, and the U.S. of 1960, could all design and produce a viable 1960 Ford. That level of achievement would be possible for them. Since they could not design and produce a viable space shuttle, it is clear that the design of the space shuttle is at a higher level than the design of the Ford. And since one level of design can be achieved while the other cannot, it is obvious that the concept of "design" has a basis in reality. It is not something we imagine or read into things.

THE GENIUS OF GENES

One of the oldest and most prestigious scientific associations is Great Britain's Royal Society. In the early 1980s, *Omni* magazine asked members of the Society to list the five most "sensational" scientific advances of the 1970s. The following is the journal's analysis of their replies:

The most frequently mentioned paper in the biological sciences was that by Fred Sanger and his colleagues at Cambridge, England, wherein they described the entire sequence of nucleotides, or "words," in the DNA of a virus, PhiX-174 (*Nature* Vol. 265, 1977, p. 687).

This achievement marked the first time ever that the complete chemical "blueprint" of a living organism had been unraveled and followed, shortly after Dr. Sanger's group and a second team working under Dr. Walter Gilbert had improved methods for reading DNA sequences. An extremely simple life form, PhiX-174 proved to contain 5,375 words. Grouped into sentences—genes—they specify the composition of a virus particle when it replicates, and indeed they control all its functions ... Quite a perplexing revelation from this work was that the genes actually overlap.

Like a telegram with no spacing, the coded message read entirely differently, depending upon whether one began with the first, second or third letter. The fact that three messages were contained within one seemed to some researchers artificial or contrived, prompting Drs. Hiromitsu Yokoo and Iairo Oshima to revise the theory, first suggested by Dr. Francis Crick and Leslie Orgel (*Icarus* Vol. 19, 1973, p. 341) that life on earth began from organisms that were sent here billions of years ago by extraterrestrial civilizations that decided to "seed" other planets. The Japanese scientists suggested that the gene sequence PhiX-174 might contain messages, or signals, that are as yet uncoded. In their line of reasoning, such overlapping messages would be a highly economical way to send information through vast tracts of space.

Thus, the most sensational biological discovery of the 1970s was that DNA, the "chemical blueprint" of a live form was so "contrived," it exhibited such a high level of design and complexity, scientists were forced to conclude that the DNA *had* to have been produced by intelligence. The scientists would not call that intelligence "G-d" but rather a mysterious civilization of "seeders" far away in outer space.

Seeding theory observes design in nature, namely the complex code of DNA, and it notes, with minor exceptions, that this code is the same for all living things. This sameness itself suggests a single source for life on earth. Still, rather than say that G-d is life's single source, seeding theory says that the single source was a higher civilization, in another solar system, fearing extinction, which therefore sent frozen bacteria out into space in spaceships, with one ship eventually reaching earth. This was the theory of Crick and Orgel.

After seeing PhiX-174, the best view of DNA to date,

Yokoo and Oshima went one step further. The higher civilization which sent the bacteria did not necessarily feel itself in danger of extinction, they said. These beings simply desired to send a message, a live one, at that!

CHAPTER FIVE

THE SOURCE OF DESIGN IN NATURE

O ne thing can be said for the scientists who support "seeding" theories. At least they realize that design is design, whether it appears in a chair, a machine or a DNA molecule. DNA is something alive, but Yokoo and Oshima did not let that stop them from concluding that the DNA had to be a product of intelligence. Crick and Orgel, the inventors of the "seeding" theory, also attributed the design in living things to intelligence. That is why they proposed their theory—to provide an intelligence other than G-d.

But to attribute design in nature, such as found in PhiX-174, not to G-d, but to a mysterious, strangely motivated civilization in outer space is simply an evasion. How did the seeder civilization itself come to be?

And so, this recent episode in the history of science is yet another example of normally logical minds malfunctioning when faced with facts pointing to a Creator. Seeding theory and its Japanese revision confirm that the "I do not know" regarding G-d stems not from insufficient data or lack of logic and reason, but from emotional and psychological terror as well as personal and social bias.

For many people, especially scientists, G-d is "too hard to take." Quality evidence of G-d is readily available, but it causes "irritation." Subconsciously, people blot out the evidence in order to avoid this irritation. Thereby, their own bias keeps them from knowing an important and vital truth.

LOGICAL ATTRIBUTION

Let the issue be clear. By what force or agency does design come into being? Frederick Ferre, in *Basic Modern Philosophy of Religion,* [16] states the answer forcefully:

> Suppose a case of books filled with the most refined reason and exquisite beauty were found to be produced by nature; in this event it would be absurd to doubt that their original cause was anything short of intelligence. But every common biological organism is more intricately articulated, more astoundingly put together, than the most sublime literary composition . . . Despite all evasions, the ultimate agency of intelligence stares one in the face.

BIONIC SNOWFLAKES

Some people may insist that design can indeed arise without a designer, however illogical it seems, and call to evidence the snowflake or crystal. These have design, and this design apparently comes about by itself, spontaneously. But this is arguing in circles. The designs of snowflakes and crystals demonstrate that they *are* the results of intelligence.

As noted above, scientists are now saying that even atoms probably did not come into being without "guidance." Furthermore, the argument stating that the force behind design in natural objects is not intelligence, with snowflakes and crystals as evidence, can easily be dismissed without resort to the recent findings of science.

The "snowflake argument" is refuted by "bionics," a not-so-new scientific discipline based on mimicry. Bionics practitioners, one of whom was Leonardo da Vinci, closely examine the design of live objects, hoping to employ the designs as models for tools and machinery that will be useful to man. Hearing devices are patterned after the ear, seeing devices after the eye. According to bionics, anyone building a ship should first look at the design of a fish. Builders of airplanes should study birds.

According to bionics, a discipline well versed in the general concept of "design," a distinction exists between the design in snowflakes and crystals and the design in something alive. Specifically, the snowflake and the crystal are not in the category of a designed "system." Rather, a "system," as defined by Garardin[17] is

a collection of components intended to perform some function. These components must be put together in a special way. If they are simply assembled at random, they cannot be said to form a system, but they are simply a structureless mass which cannot accomplish anything.

Components assembled in order, however, do not necessarily form a system either.

A crystal, for example, is a wonderful organization of atoms, but cannot be said to form a system, as its arrangement is not intended to perform any function; it is an end-product in itself and does not change unless [if] by some accident.

By contrast, a living cell is changing continuously, nourishing itself and reproducing itself in order to perpetuate the life within it. The living cell is a true system because it has a purpose; the crystal has no purpose and cannot be called a system. The arrangement of the components within a system cannot be random, because a system always has a goal.

In other words, the design that is inherent in a snowflake or a crystal is not nearly as intricate as the design that is in a "system" such as a cell. The design in a snowflake or crystal is not a design allowing the object to change, to produce or to accomplish some function.

Granted, the snowflake and the crystal exhibit design, just as the *2001* monolith exhibited design. However, in comparison to the level of design inherent in "systems," the level in the snowflake and the crystal is quite low. Therefore, even if the design in snowflakes and crystals arises spontaneously, without intelligence, it does not follow that the much more intricate design of living systems could arise spontaneously.

Consequently, when living systems exhibit "sublime" and "astounding"elements of design, "the ultimate agency of intelligence stares one in the face," because in all of human experience, the only force able to produce this level of complexity is intelligence. Only an intelligent force can be behind the design in living systems.

Nevertheless, cognitive dissonance can cause people to think otherwise.

CHAPTER SIX

THE EVOLUTION
REVOLUTION

We have shown that nature undeniably exhibits an intelligent order, which naturally implies an intelligent Creator. This reasoning is so powerful that those who fear its logical conclusion, or who fear imagined conclusions, muster all their imagination to concoct alternative explanations for the obvious design in nature. The conventional attempt is known as the theory of evolution.

EVOLUTION

In the nineteenth century, Charles Darwin theorized that living systems came into existence spontaneously, without the intervention of intelligence. He asserted that living systems owe their existence primarily to chance and "natural selection." It is important to realize that Darwin had no proof at all, yet this theory of evolution became a tremendously popular idea the moment it was proposed. And although scientists recognized at the outset that Darwin's theory had weaknesses, anyone who dared hint that Darwin might have been wrong brought upon himself heaps of scorn from his colleagues. The theory of evolution quickly appeared in the textbooks as fact, rarely as theory.

These days, Darwin's particular ideas have fallen into deep disfavor. Not only are there few facts to support Darwin, there are actually many facts that refute him. Nevertheless, though his original theory is no longer taken seriously, the general idea of "evolution" remains as popular as ever. It still survives in many forms. That is, many scholars have published telling criticisms of evolutionary theory, from many perspectives, but rarely is anyone bold enough to discard evolution entirely. For example, Nobel Prize winning chemist Dr. Harold C. Urey[18] admitted:

All of us who study the origin of life find that the more we look into it, the more we feel that it is too complex to have evolved anywhere.

Yet, he added an incredible "but":

We believe as an article of faith that life evolved from dead matter on this planet. It is just that its complexity is so great, it is hard for us to imagine that it did.

It is curious, indeed, that a theory which must fall back on "faith" is entrenched in the minds of masses. It is stranger still that such a theory has consistently won the hearts of scientists. Usually, when a theory which claims to be scientific is found to contain even a slight flaw, the theory is scrapped. Yet the theory of evolution with its different versions, always bearing major flaws, has managed to remain a sacred cow. Criteria and standards rigidly adhered to when it comes to other theories are not adhered to at all in the case of evolution. This has been the case since Darwin, and it continues this way today, for one reason: On the issue of whether design in nature is the work of a designer, absolutely no middle ground is available. The answer is either "No, it is not" or "Yes, it is."

The puzzling longevity of Darwinism, despite its many flaws, is striking confirmation for the thesis of this book. Without evolution, man is "stuck" with G-d. Subconsciously and consciously, scientists, journalists and others cling steadfastly to evolution. Because the idea of evolution allows people to entertain the notion of a universe without G-d, evolutionary theory survives and flourishes in many versions, and objections to evolution are brushed aside with scorn.

At the Alpach Symposium, one of the recent symposia where problems with Darwinism were discussed by biologists, statisticians, and others, "sociological" factors were said to explain the theory's survival.[19]

> I think that the fact that a theory so vague, so insufficiently verifiable and so far from the criteria otherwise applied in "hard" science has become a dogma can be explained only on sociological grounds.

Sociological factors have admittedly played an important role in keeping *all* evolutionary theories alive. But they are only one part of the general *psychological* factor of cognitive dissonance.

Evolution is a sacred cow not simply because socially it is "in" to believe in it. Evolution is what it is because to many people, especially scientists, the theory's downfall would be psychologically irritating and therefore unacceptable.

Indeed, in *Evolution from Space*,[20] Britain's most eminent astronomer, Sir Fred Hoyle, documents "howling" problems with the theory of evolution and concludes that the theory survives only because

> [It is] considered socially desirable and even essential to the peace of mind of the body politic.

Like Vonnegut's favorite character in *Breakfast of Champions*, the scientific and lay world is stricken with sudden terror when faced with the possibility of "meeting" our Creator. The implications are so jarring, people cannot think straight. Let us show manifestations of this phenomenon, especially with regard to scientists and journalists.

THE MUTATION FASCINATION

As we mentioned, Darwin began the process by proposing that the agents behind the live "systems" in nature are blind chance, inheritance of acquired characteristics and natural selection. The science of genetics and heredity, however, contradicted this idea, and upon the discovery that cells undergo mutations, scientists finally proposed "synthetic theory." According to synthetic theory, the agent in evolution was the mutation.

What exactly is a mutation? All life forms are made up of cells. Within the nuclei of all cells are genes, each supposedly a compact, computer-like code of information about a certain trait of the whole life form. One sheep gene, for example, might carry within it the code for thickness of wool. Other genes would be for other body traits.

The cells of plants and animals constantly renew or reproduce themselves by means of "cell division." In cell division, a plant or animal's entire genetic code is reproduced. A copy is made. In reproduction of a whole life form, the genes of a male mix with those of a female, and the result is offspring with traits from both parents.

When, every now and then, at random, the genes do not copy correctly, the result is a mutation.

If the mutation occurs in reproduction cells, it can manifest itself in offspring. According to synthetic theory, offspring with beneficial or "positive" mutations are "selected" by the environment to survive. Over billions of years, long chains of beneficial mutations in successive offspring are said to lead not only to changes within a species but also to the formation of entirely new species.

There are two major problems with this theory. First, if a copying mistake occurs in a particular reproductive cell, in the male, for example, the male cell might not be able to unite with a female cell at all, and no resulting offspring would appear from the union.

Second, even when mutated male cells do unite with female cells, the resultant offspring are almost always unhealthy. They are weak, diseased and often deformed. In other words, copying mistakes in reproductive cells either prohibit reproduction or generate offspring that natural selection would eliminate. Synthetic theory is based on a faith that mutations moved life "forward." The fact is, however, mutations, in an overwhelming number of cases, are either steps back or they are not steps at all. Indeed, Nobel Prizewinner Dr. Ernest Chain[21] has declared:

> To postulate that the development and survival of the fittest is entirely a consequence of chance mutations seems to me a hypothesis based on no evidence and irreconcilable with the facts. These classical evolutionary theories are a gross oversimplification of an immensely complex and intricate mass of facts, and it amazes me that they are swallowed so uncritically and readily, and for such a long time, by so many scientists without a murmur of protest.

A second telling blow to synthetic theory involves fossils. If gradual, consecutive, positive mutations in offspring are the agents by which species emerge and evolve, the slow steady progress of evolution should be recorded in rock for all to see. There should be fossils corresponding to each "rung" in evolution's "ladder." Concerning any species alive today, it should be possible to find fossil records not only of distant ancestors, but also of the many "transitional" forms in between.

When Darwin was alive, he admitted that the fossil record did not "yet" provide evidence that evolution has actually taken place. The record was packed with "missing links" in the alleged evolutionary chain. The "ladder" lacked rungs. Darwin was confident, however, that if paleontologists would dig in the right places, the "missing links" would certainly turn up.

However, a century of digging since then has not uncovered the fossils to support Darwin's claim. Paleontologists have devoted entire careers to finding fossils that would show gradual changes in life forms over time. The search has been in vain. Indeed, for evolutionists, not being able to find the "missing links" in the alleged chain of development is "a professional embarrassment." Writes Hoyle:

> It is not hard to find writings in which the myth is stated that the Darwinian theory of evolution is well proven by the fossil record. But one finds that the higher the technical quality of the writing, the weaker the claims that are made.[22]

In textbooks admitting the problems of Darwinism, the blame is constantly placed on "imperfections" in the fossil record:

> Yet, if one persists by consulting the geological literature, the truth eventually emerges. The fossil record is highly imperfect from a Darwinian point of view, not because of the inadequacies of geology, but because the slow evolutionary connections required by the theory did not happen. Although paleontologists have recognized this truth for a century or more, they have not been able to make an impression upon opinion.[23]

Here we have outlined but two of many reasons why the synthetic theory is invalid. For more details, readers should refer to sources listed in the appendix of this book. Now, let us examine what other alternatives scientists are offering to explain the development of life.

HOPEFUL MONSTERS

First and foremost, is "punctuated equilibrium," a take-off on the "hopeful monster" theory of evolution first proposed by Richard Goldschmidt.

Goldschmidt suggested that changes in animal and plant life do not occur slowly and gradually over long stretches of geologic time, as evolutionists have held in the past. He proposed that evolutionary changes occur in "leaps," quickly and suddenly. Small copying mistakes in genes produce large "monstrous" changes in offspring.

Goldschmidt agreed that almost every monster would be eliminated by natural selection. Occasionally, however, one of these freaks would survive and, thereby, a new species would be born! Goldschmidt dubbed these new arrivals "hopeful monsters."

Punctuated equilibrium, building upon Goldschmidt's "leap" theory of evolution, asserts that geologic time generally maintains equilibrium, with life forms not changing at all. Genes copy correctly. These vast stretches of time, however, are "punctuated" by brief periods of drastic change.

Very neatly, punctuated equilibrium explains away the embarrassing gaps in the fossil record. There are no transitional fossils because there never were any transitional creatures! New life forms literally snapped into existence. Presto! Mind you, the copying mistakes in the microscopic genes had to have been small. Otherwise, the offspring would not have been born at all. Nevertheless, the small mistakes somehow produced wholesale changes in offspring. This is the latest version of the theory of evolution.

Newsweek magazine (March 1982) writes:

> For all the excitement it has generated, punctuated equilibrium still smacks of heresy to many scientists. It does not explain what many regard as the crucial point: how and why a new species springs up.

In other words, what is the stimulus for the leaping? What makes the genes copy incorrectly in certain eras and not in others? Also, why should genes suddenly *stop* making mistakes in copying? Finally, even if gene copying mistakes generated a male "hopeful monster," how could the new species survive if at the same time there was not a similar female able to mate with him?

The proponents of punctuated equilibrium have no answers to the crucial questions.

Newsweek, however, saw fit to run a cover story about punctuated equilibrium and its most vocal proponent, Stephen Jay Gould. Gould's silence on the crucial points is not important in *Newsweek's* eyes. *Newsweek* makes no mention at all of Goldschmidt or of mutations, or of the overwhelming odds against the birth and survival of "hopeful monsters." Gould admits that his theory assumes the existence of millions of such undetected monsters, yet *Newsweek* is not the least bit skeptical.

Just as scientists have rushed to accept other theories of evolution uncritically and readily, in total disregard of their obvious shortcomings, *Newsweek* rushes to accept punctuated equilibrium.

THE PANDA BEAR THUMB

Gould gives us no reason to believe in his theory, but he does offer us a reason to disbelieve in G-d. He presented his argument in an essay *Newsweek* falls for, head over heels. Everyone agrees the panda's thumb is "beautifully suited to the panda's sole occupation, which is peeling leaves off tasty bamboo shoots." Gould feels it is important to point out, however, that the panda bear thumb is not very much like the human thumb, or the thumb of a monkey. It is not an independent "digit." Rather, the panda bear thumb is an extension of the panda bear wrist bone. Gould says flatly, "If G-d had made the panda, he would have done a neater job . . . [The thumb is] inelegant . . . [and] its only virtue is that it works."

If Gould had been in charge, the thumb would have been made differently. The thumb works perfectly, but Gould has his own professional opinion, his own subjective standard of what anatomy should be, and suddenly there is no G-d. This is science?

Newsweek, though, is taken in. It prints an illustration of the panda bear thumb, with the caption, "Would G-d have done such a messy job?" *Newsweek* calls the arguments of those who believe in G-d "unsophisticated," while the panda bear argument and Gould's theory are "subtle," with "attractions to the modern mind."

Like so many evolutionists before him, Gould would have us ignore the many fundamental flaws in his theory. "Evolution is a fact," says Gould, "like apples falling out of trees." *Newsweek* nods approval, saying that we can be sure that it will go on.

The bizarre behavior of Gould and *Newsweek* cannot be explained simply by "sociological" factors, the desire to be "in." Great psychological pressure must be exerting itself as well. Cognitive dissonance about G-d can stop people from thinking rationally, even people who are highly intelligent. Gould and *Newsweek* are victims of sudden terror. With the demise of the synthetic theory of evolution, they are scrambling. For whatever reason, they are so afraid of "meeting" G-d, they cannot think straight.

LIFE ORIGINS VS. LIFE DEVELOPMENT

We have been discussing theories of evolution because when the question is posed, "What produced design in nature?" people will often reply, "Evolution," not G-d. Technically, however, the turf of evolution is life's development, not life's origin. Indeed, most scientists distinguish between life development theories and life origin theories, disagreeing with those who maintain that "origin" is included in development.

Darwin, in his own writings, premised his theory on the assumption that a protein compound could be formed chemically, by chance, "in some warm little pond, with all sorts of ammonia and phosphoric salts present . . . ready to undergo still more complex changes."[24] In other words, even with Darwin's theory of life's development, a separate theory of life's origin must be provided to explain life on earth.

So it is with every theory of life's development. To maintain that the source and force behind design in nature is not Divine Intelligence, one must present not only a viable G-dless theory of life's development, one also must present a viable G-dless theory of life's origin. Having seen that G-dless life development theories are seriously flawed, but flourish because of dissonance, let us turn to the G-dless life *origin* theories, to see if the situation differs at all. If, in fact, there is no viable G-dless theory of life's origin, then we can again expect some bizarre behavior and weird thinking on the part of scientists and journalists.

CHEMICAL SOUP

In the nineteenth century, Darwin and almost all other scientists agreed that "simple" life arose on earth spontaneously, out of early Earth's lifeless chemical soup. After vast stretches of geologic time, the appropriate chemicals were said to have shuffled themselves together into "primitive" life, by chance. This is the chemical soup theory of life's origin.

Like Darwinism and evolution, the chemical soup theory is full of holes.

First of all, even the most "primitive" life is quite complex, as Britain's Sir Fred Hoyle has illustrated, using enzymes. Enzymes are a class of chemicals essential to life. There are about two thousand different enzymes, and each one has its own distinct structure. The origin of the entire class is unknown.

According to Hoyle, the chance of obtaining all two thousand enzymes randomly is one in $10^{40,000}$ "which is about the same as the chance of throwing an uninterrupted sequence of fifty thousand sixes with unbiased dice." These calculations say nothing of the odds that chance produced the "programs" by which cells divide and organize themselves. "These issues are too complex to set numbers to," says Hoyle. "Likely enough, however, the chances of such complexities arising from a soup that can proceed only by trial and error, are still more minute."

A second "hole" is, according to scientific findings, that Earth's first "primitive" life appeared early, not late, in the Earth's history. Paleontologists have found fossil remains of live cells almost as old as the solar system itself. Therefore, even if chemistry and blind chance could have "rolled all the sixes" and produced something alive, all the failures to roll correctly would have taken time—lots of it. Between the start of the chemical processes and the advent of life, there simply was not enough time for chance to work.

As a result, says Hoyle, "If one is not prejudiced either by social beliefs or by scientific training,"[25] the chemical soup theory "is wiped out of court. [It is time someone] blew the whistle."[26]

A third "hole" in the chemical soup theory has been discussed by a chemistry professor, Robert Shapiro, in his widely acclaimed book, *Origins*. According to the soup theory, inorganic chemicals in Earth's early atmosphere are said to have shuffled themselves together into something alive. The original stimulus is said to have been an electric spark, or some other outside source of energy, interacting with Earth's atmosphere.

The interactions are said to have led to the formation of organic compounds, which accumulated and made their way into the soup of Earth's primitive oceans. These organic compounds were the so-called "building blocks" of the first life, having somehow, by means of unexplained processes, organized themselves into something alive. Shapiro[27] discredits the experiments that claim to prove the possibility of this scenario. He proves that these experiments do not actually simulate what science knows about Earth's early atmosphere. Moreover, they produce nothing that even resembles a living being, and hardly any of the organic molecules that are essential "building blocks" for simple life.

Even so, the experiments are widely acclaimed as being conclusive proof of the soup theory and have been "incorporated into high school and university biology and geology texts, and featured in museum displays."[28]

According to one textbook, such "Miller-Urey experiments" produce, from a simple mixture of gases, "organic compounds . . . including representatives of all the important types of molecules found in cells, as well as many not found in cells."

"That statement," writes Shapiro, "is simply incorrect," and the more one looks at what the Miller-Urey experiments produce, the clearer it becomes that the soup theory is simply a myth, with no scientific basis at all.

As to why textbooks contain such distortions, and why the soup theory has had such great impact despite its shortcomings, Shapiro writes that the answer "involves psychology and history more than chemistry." The soup idea, when it was first proposed (in the 1920s), "filled a near vacuum" that had arisen in the halls of science since Louis Pasteur, who had disproved the widely held notion of spontaneous generation (the notion that life can arise out of non-life).

Pasteur had demonstrated that living beings arose only from earlier living beings. How then did the first life arise?

In the absence of a viable scientific answer, those needing a solution could turn only to religion. To some scientists, particularly those defending evolution from attack by fundamentalists, this situation was unacceptable. The most obvious remedy was the revival of spontaneous generation in some form, with the added provision that it required conditions that were present long ago on earth, but not now.

In addition, the thought arose that the formation of an entire bacterium might not be necessary. To start life, it might suffice if some smaller part of a cell, a protein or even a bit of gel-like protoplasm, came into being.[29]

The Miller-Urey experiments produce hardly any of the necessary "smaller parts," and the experiments say nothing at all about how the many "smaller parts" shuffled themselves together to bring the first life into being. Nevertheless, the Miller-Urey experiments have been passed off as "proof" that the chemical soup theory of the origin of life is true.

And so, as we suspected, there does seem to be a parallel psychological factor behind *development*-of-life theories and *origin*-of-life theories. When the "synthetic" theory of evolution, the popular theory of life's development, was found to have no basis in scientific fact, scientists who normally think rationally started to scramble. Without something to "fill the vacuum," scientists came face to face with the idea of G-d, and this terrified them. Outlandish theories such as punctuated equilibrium were proposed, and the media applauded the new theories!

And so, too, after a "vacuum" was created by Pasteur, scientists in the 1920s scrambled to accept the soup idea as a viable theory for the *origin* of life (as opposed to its *development*). In 1952, when the first "spark" experiments were performed, and the results were reported, scientists and the media triumphantly announced that the soup idea had been proven experimentally. Textbooks and newspaper articles about the experiments began to trumpet tremendous distortions and outright lies.

It is a problem of "psychology," writes Shapiro. It is cognitive dissonance. Shapiro observes that the reason scientists have fed the soup idea to the public for so many years is that it serves to fill that awful "vacuum." Scientists and the media *want* the soup idea to be true, by hook or by crook. Rather than accept the "religious" idea about the origin of life, scientists and the media work to dress up a myth to make it look like science.

In the end, students and laymen are misled to believe that science has found the answer to the origin of life, an answer which eradicates the need for G-d as the solution.

It has been a great hoax, writes Shapiro, because in truth, chemical soup theory has no foundation, and scientists do not have the foggiest notion about how life began.

SEEDING

As we wrote earlier, Britain's Fred Hoyle rejected Leslie Orgel's and Francis Crick's proposal of the seeding theory to explain the origin of life. Because they saw that the PhiX-174 DNA is so "contrived," Crick and Orgel did declare it must be the result of intelligence. But, they asserted, DNA is not the work of G-d but of a "seeder" civilization somewhere in outer space.

This was their alternative to accepting the concept of a Creator, while at the same time steering away from the discredited idea of life having arisen spontaneously out of Earth's "soup."

The reason Hoyle rejected the Crick-Orgel idea is that it did not explain the origin of the original seeder. Furthermore, through the calculations he made using enzymes, Hoyle concluded that the same odds precluding the possibility that life could have arisen by chance out of *Earth's* chemical soup preclude the possibility a "seeder" have arisen by chance in *any* chemical soup, anywhere in the universe, regardless of the soup's size.

The odds against chance and chemistry being responsible for life, Hoyle wrote, "are essentially just as unfaceable for a universal soup as for a terrestrial one."[30] In other words, if Earth's chemical soup could not have generated life without the intervention of intelligence, neither could the chemical soup of the entire universe. Hoyle added:

> No matter how large the environment one considers, life cannot have had a random beginning. Troops of monkeys thundering away at random on typewriters could not produce the works of Shakespeare, for the practical reason that the whole observable universe is not large enough to contain the necessary monkey hordes, the necessary typewriters, and certainly the waste paper baskets required for the deposition of all the wrong attempts. The very same is true for living material.[31]

Hoyle saw that if one does not introduce the ingredient of intelligence, the chemical soup theory of life's origin is a fallacy:

> Biochemical systems are exceedingly complex, so much so that the chance of their being formed through random shufflings of simple organic molecules is exceedingly minute, to a point where it is no different from zero ...

For life to have originated on earth it would be necessary that quite explicit instructions should have been provided for its assembly.[32]

In other words, if life on earth arose from earth's own materials, the process must have been guided by an outside intelligence.

In fact, Crick and Orgel themselves brought up the entire idea of seeding only to "increase public awareness" and "awaken" people to the demise of the chemical soup idea. Crick himself confided this to Professor Robert Shapiro, mentioned above, who writes that Crick told him:

> We thought of this theory, but we're not completely sold on it . . . The object is to give the intelligent person an idea of what the problem really is, and this is just a tag to hang it on . . .
>
> Everybody, as they say in the state of California, can relate to certain ideas, and things like coming on an unmanned rocket, or even bacteria, they think they can relate to.[33]

In other words, scientists had put up so much "smoke" about the chemical soup theory to make it look valid that they had succeeded in bamboozling the public into believing in it. The soup idea had become so entrenched that scientists who knew it was a hoax felt compelled to propose the alternative theory of seeding in an attempt to dissipate the smoke!

Crick and Orgel probably observed that they could not propose the creation of life by G-d as an alternative to the bankrupt soup idea because people already were indoctrinated to G-dless theories of life's origin and development. Just as someone indoctrinated to believe in the pervasive propaganda of Big Brother could not possibly accept the truth about the stars, a society believing so strongly in soup and "evolution" could not possibly accept the idea of G-d.

When the chemical soup theory proved itself to be a hoax, and a "vacuum" was created, Crick and Orgel sought to fill the vacuum with an alternative theory of life's origin, but people had become so brainwashed that the only thing they would "relate to" was unmanned rockets and imported bacteria! Seeding! When Crick and Orgel put forth the seeding proposal to the scientific community, they did not believe in it themselves, and for good reason.

But what happened? Seeding theory was taken seriously—it was enthusiastically applauded and it rapidly gained a large following! People, especially scientists, grabbed onto it for dear life. Indoctrinated, having invested so much in the G-dless belief system, they accepted even the outrageous, simply because it allowed them to continue in their own ways. Right or wrong, the idea of G-d is was like a lion, ready to pounce and hungrily devour them. They picked up any stick readily available to keep the lion away.

According to Hoyle, a highly respected and admired scientist, the Ultimate Agency must be G-d, a Supreme Being whose powers and intelligence cannot possibly be fathomed by the mind of man.

Hoyle preserved the essential idea of a "seeder," but admitted that the seeder must owe *its* origin to a Supreme Being.

Hoyle's ideas attract the attention of *Newsweek*, but in *Newsweek's* article, the word "G-d" does not appear even once. Tongue in cheek, *Newsweek* (March 1982) says only that Hoyle

> has actually performed the improbable feat of re-inventing religion . . . [and has been] led to exactly the same view that seemed prevalent in the Middle Ages: that life did not arise spontaneously on earth.

Apparently, when Hoyle the scientist is led to G-d, *Newsweek* is irritated.

Shapiro, too, is irritated. Life's origin is a mystery, says Shapiro. It is a grand enigma that science may never solve, but to admit to G-d is "unscientific." Shapiro writes that even though there is no evidence that life exists elsewhere in the universe, still, to attribute the great string of "coincidences" to G-d is "unscientific." He writes:

> The claim that the universe, the earth and life were made by an undetectable Creator using supernatural powers falls outside of science. It makes no predictions that can be tested. It cannot be negated by science.[34]

Shapiro admits that science has no better theory about how the universe and life came to be, but in his opinion, it is better to live in a "vacuum" than to accept the idea of G-d.

PRESSURES WITHOUT AND WITHIN

When a person grapples with the question, "What is the source of design in nature?" or with any question that relates to G-d, the true answer is bound to elude him. From the outset, one must realize that external and internal influences will be at work to create confusion.

HOYLE'S STRENGTH AND WEAKNESS

As we saw above, design in nature had led Sir Fred Hoyle to see that there must ultimately be an intelligent Creator. It seems that he remained free of external pressures, at least insofar as accepting Darwinism and the chemical soup theory. Hoyle was not swayed by his professors, colleagues, friends or the media, all of whom had one basic opinion in these matters, which most of humanity had rushed to accept. Hoyle kept his head. To his credit, he (like Einstein) detached himself from these external influences. As a result, Hoyle was able to recognize the existence of G-d.

However, concerning issues relating to G-d, there are going be additional confusing influences from within, which we have discussed above. For example, a person might have an aversion to the idea of being a creation of a Higher Being. Man, who is himself a creative being, perceives, rightly or wrongly, that he is belittled by the idea of G-d.

In addition, acceptance of G-d can "belittle" a person simply because G-d in some ways is difficult to understand. To accept G-d is to admit that man is "small," in terms of his understanding. These different feelings of inferiority that are caused by the idea of G-d produce internal psychological pressures which can confuse a person's thinking. They can lead a person to reject the idea of G-d completely, in order to avoid the feelings of puniness.

Also, as we have pointed out, one appreciates that in a table, a clock or any other man-made object, the design in the object implies that the object has a purpose and a goal. Accordingly, if the high level of design in the human body is said to be the work of G-d, and so, too, the design of nature and the universe, then it follows that G-d must have a purpose and goal in mind for these "designs." If He did not, why create the designs in the first place? As soon as a person suspects that his ideas, thinking or behavior may not be wholly consistent with what G-d has "in mind," the perceived inconsistency creates internal psychological pressure on him to reject the idea of G-d.

At a minimum, the internal pressure confuses the person's thinking about G-d. If someone feels that his accepting G-d would make him feel "small" in some way, or it would compel him to make changes in his thinking or behavior, evidence of G-d will create "dissonance" in the person, and the dissonance can lead to irrational thinking, even if the person is intelligent. Because of such internal pressures working against the idea of G-d, the truth can be utterly lost on people, even though the evidence is overwhelming.

However, although *Newsweek* describes Hoyle as performing the improbable, having "reinvented religion," if we examine the conclusions of Hoyle, as he states them, and if we read Hoyle's own account of how the conclusions were reached, it will become obvious that *Newsweek's* reaction was quite premature. In actuality, Hoyle has not really reinvented religion. Not in the least.

Although it seems that Hoyle resisted the external pressures against accepting G-d as the Creator of the universe, Hoyle did, in fact, succumb to the internal pressures. Hoyle admits that the development of life must have been guided by intelligence and he rejects the idea that life on earth could have arisen from earth's own materials (as religion maintains) through instructions contained in natural laws of physics and chemistry.

Why does he reject the idea that earth's own materials could produce such intelligent design? Hoyle explains that this is unacceptable because from there

> it is only a short step to say that G-d created the laws, and then we are back to the special creation theory that Darwinism was supposed to have replaced.[31]

Hoyle prefers to believe that life at first arose not on earth, but extraterrestrially, somewhere in outer space, and somehow, "seeds" from that life reached earth. The "seed" materials made their way to earth after having been "programmed" or "instructed" to develop into life, in all of its forms. The designer of this "program" was a mysterious intelligence other than G-d. Because of the program, evolution on earth proceeded quickly and smoothly, without mistakes, according to the plan of the seeder.

It should be clear that Hoyle's version of "seeding" theory is no less flawed than the "seeding" theory proposed by Crick and Orgel. In fact, Hoyle's version of the seeding theory seems even more irrational. He proposes that life on earth owes its origin to a non-divine intelligence far off in space, which sent programmed raw materials to develop into every species alive on earth today, including man.

At the same time, Hoyle admits to the existence of G-d, for just as the earth's soup could not have produced even simple life without the intervention of intelligence, neither could any soup, even the soup of the entire universe. No soup could have produced Hoyle's alleged programmer, either.

That is, Hoyle recognizes the Creator and that the ultimate explanation for any alleged programmer must be G-d. As a result, the logical question is, "If Hoyle acknowledges the existence of G-d, why must a programmer be responsible for life on earth? Why can't Hoyle admit that the intelligence behind life on earth may be G-d?" Between man and G-d, why must Hoyle insert a seeder?

A THEORY WITH AN ADVANTAGE

We find the answer in Hoyle's own writings. Hoyle writes that if he is correct in his belief that life on earth is imported from some place located in the vastness of the universe, sent by a non-divine seeder, and planet Earth "is not the biological center of the universe," then there is a certain "advantage."

> The advantage of looking to the whole universe . . . is that it offers a staggering range of possibilities which are not available on earth. For one thing, it offers the possibility of high intelligence within the universe that is not G-d.[35]

Hoyle reveals his mind by saying that the idea of a "seeder" intelligence far off in space is a nice escape from the idea that G-d is responsible for life on earth. As evidence piled up in front of him indicating that life is too complex to have arisen by chance, and even a seeder must have a creator, Hoyle writes, "a monstrous spectra kept beckoning." To Hoyle, the very notion that everything must trace back to G-d is "monstrous."

Again, Hoyle may be very wrong in thinking that the idea of G-d is "monstrous," but because he thinks that way (i.e. because the idea of G-d produces so much cognitive dissonance in him), his normal thought processes are totally disrupted. Because the idea of G-d is so "monstrous" to him, his real interest is not to find a theory which offers truth. Rather, Hoyle's most pressing concern is to "escape" and to create an "advantage." As a result, Hoyle imagines a "programmer," a seeder, and by inserting his seeder into the universe, Hoyle seeks to break the "monstrous" direct connection between man and G-d. For some reason, the direct connection is a very threatening prospect to him.

And so, according to Hoyle, the Ultimate Source of life in the universe is G-d. Also, G-d is transcendent, with powers and intelligence even the alleged seeder cannot grasp. Finally, Hoyle believes that G-d is not involved with anything in creation. That is, G-d's involvement is ruled out by G-d's transcendency. Man should not seek to find G-d in the universe, for G-d is "beyond" the universe, "above" it, and not involved. Hoyle says that man would do better to search for clues to the whereabouts of the seeder.

But a question is nagging: Once Hoyle sees G-d's intelligence and that G-d's powers cannot be fathomed, why is he so certain that G-d is powerless when it comes to involvement? Why doesn't Hoyle realize that G-d has the power to be involved, despite transcendency? Why is Hoyle so certain that G-d is detached from us?

The answer is clear. In the mind of Hoyle, the idea of G-d is "monstrous." As a result, Hoyle has interests which distort his thinking. Therefore, although *Newsweek* says otherwise, there is nothing about Hoyle that reinvents religion. No, Hoyle's theory has its "advantage." Hoyle regards G-d as "monstrous," and suddenly and necessarily, G-d is detached. A ghostly seeder appears. Man escapes. Hoyle is in the grips of cognitive dissonance.

The external factors influencing most people to regard creation as myth, and evolution as unquestionable fact, include our educational institutions as well as the mass media. As Hoyle wrote:

> Once the whole of humanity becomes committed to a particular set of concepts, educational continuity makes it exceedingly hard to change the pattern.[36]

Nevertheless, schools and textbooks are not the sole partners of the media. Friends, family and co-workers exert their particular pressures, as well. When all say that the world is flat, it takes courage and strength to say no, the world is round. When all believe in evolution, who has the fortitude to take a stand for G-d? The pressure to conform to the opinion of the majority is formidable. Unless a person can free himself from the external and internal pressures, the truth about G-d will elude him.

Now let us summarize. We have been investigating different possible sources for the design that is manifested in nature. Theories of evolution, old and new, fail miserably in their attempts to depose intelligence as the logical source, and soup and "seeding" are not viable theories of life's origin. Furthermore, no theory about the origin or development of life even addresses the many "coincidences" which led to the design in the inorganic realm of the universe.

Nevertheless, a person is under great social pressure to accept evolution, to accept the soup idea, and to accept any theory of life which does not mention G-d.

G-dless theories are presented and accepted as fact. A tremendous amount of brainwashing has been going on. As a result, a person's ability to have the intuitive appreciation that G-d exists is impaired. In other words, although seeding, soup and the different theories of evolution are all hoaxes, each is a block preventing people from seeing G-d in nature.

We have shown that "design" is not something people imagine. It is not subjective or something people "read into." That took care of one block between man and G-d.

Then we showed that although live things are different from machines such as the space shuttle, the design in live things is no less real than the design in machines. That took care of a second block.

All that is needed to remove the block presented by anti-G-d pseudoscience is straight information and clear thinking. The arguments and sources presented here, we feel, are clear indicators in the path toward clear thinking.

CHAPTER SEVEN

BELIEVING WITHOUT SEEING

I t might be argued that it is not cognitive dissonance which keeps people from recognizing G-d in nature. Rather, G-d's existence is not obvious because people are lacking certain necessary experience, without which conclusions about G-d cannot be drawn.

In other words, a clock in the desert, or a black slab such as the *2001* monolith, is an object of which people have prior experience. People have seen these objects and know how they are made.

Live objects, however, are much more complicated than inanimate objects and people have no prior experience of the purposeful manufacture of anything alive. Perhaps this, not cognitive dissonance, is the reason people fail to see intelligence behind the design in nature's live objects.

When faced with a baby or a potted plant, people fail to conclude that the design is the work of a Designer, not because the conclusion makes them feel uncomfortable, but because they are lacking the necessary prior experience, the mind simply cannot make the connection.

This, it may be pleaded, is why film-viewers saw the right angles and smooth surfaces of the slab in the *2001* film and easily accepted the conclusion as logical that the slab must have been designed and manufactured purposefully by some intelligent entity.

The slab was a non-living entity, as is a table or a clock. People have prior experience of tables and clocks being manufactured by intelligence, and intelligence was therefore the logical explanation for the slab's (minimal) complexity. But for live objects there is no such prior experience.

The response to this begins with a simple observation. Let us say a person is seated in a room with a table, a clock, a potted plant and a baby.

We ask him, "Which objects in this room were designed and manufactured purposefully?"

He answers, "The table and the clock."

Is he certain?

Of course he is.

Without question he is able to conclude that both the table and the clock are products of design.

However, confident certainty is not based on personal experience of tables and clocks actually being made. The certainty is the same even after seeing only other non-living objects being made purposefully. A person is *told* that tables, clocks, and other familiar household objects are made purposefully. Here, therefore, the source of absolute knowledge is not direct experience. It is very easy to be absolutely sure that an object is a product of design inherent in the object, even if actual experience of the design is absent.

In fact, this principle is already operative in the minds of those astronomers who still hope that there might be intelligent life somewhere other than on earth. This is their thinking: If somewhere in the universe there is non-human intelligent life, perhaps there are non-human civilizations which are capable of space travel. If so, maybe non-human creatures will someday approach the earth in spaceships. Now, if these non-human space travelers enter our solar system and glance at planet Earth, if they are too far away, they will not be able to pick out anything on the Earth's surface which would indicate that there is intelligent life in existence.

According to a noted astronomer Carl Sagan, however, if the aliens came sufficiently close, they would find "unambiguous" and "convincing evidence of intelligent life on earth," as soon as they sighted things such as city crossroads, neat suburban housing or contour plowing and planting. Sagan writes[37] that without ever having seen the streets being paved, the suburbs being built or the fields being plowed and planted, the aliens would *infer* that intelligent life exists here. Direct experience of the design would not be necessary.

Really, the thinking of astronomers is this: If intelligent life exists in the universe on some other planet, maybe this alien's intelligence is on a par with man's. If so, it would probably be possible for man to distinguish features on another planet's surface similar to city crossroads, suburbia and patchwork plowing. Therefore, if man ever sets out into space looking for non-human intelligent life, man will not have to touch down on every single planet.

Instead, man could gaze down on the planets from pre-determined altitudes and *infer* intelligence. Again, all that would be seen would be an effect, i.e. a final product. Actual experience of the cause of the effect would not be necessary. Therefore, just as human space explorers could infer that intelligence is behind designs on other planets, you can infer intelligence is behind design in nature.

From many perspectives, it is clear that direct experience is not essential for sound decision-making. Imagine a courtroom, for example, where a murder trial is drawing to a dramatic close. On the basis of eyewitness testimony, the jury is pronouncing its verdict.

"We find the defendant . . . guilty," says the foreman.

Suddenly, the defendant jumps up and cries, "How can you say I'm guilty? You weren't there! Without personal experience you can't know anything!"

His cry is ignored, and rightfully so. No judge or jury regards lack of personal experience as an obstacle to sound decision-making.

Or let us say you are visiting San Diego for the first time, and you want to see the famous zoo there. You ask a policeman where you can catch the bus, and you stand where he tells you. Along comes a bus marked "San Diego Zoo." Do you get on? How can you? After all, you have no personal experiences of this bus actually reaching the zoo. You do board the bus, though, don't you?

Or let us say you work in a store, and a buyer comes in who wants to pay for something with a personal check. You have not been to the fellow's bank to actually see that there is cash there to cover the check, but this lack of personal experience does not stop you from selling to him, does it?

People are willing to make countless decisions in daily life, and act in concrete ways, even in the absence of firsthand experience. In fact, if people were *not* willing to decide and act without seeing things firsthand, banking, commerce and many other daily procedures would be stalemated!

If one can accept a check or one can board a bus without firsthand experience, and one can sit on a jury and evaluate the relative merits of a case, or look at a chair and know intuitively that it was designed, then one should be able to decide about the design of living things in nature (as Yokoo and Oshima decided) even though the actual designing was never personally witnessed.

People constantly act and decide in the absence of firsthand experience, so why do they shy away from doing so when the issue is design in nature?

Observe the force of magnetism. Magnetism itself is not detectable by any human sense organ. Even though the human being has no "ear," "eye," or other sense organ able to detect magnetism, he is nevertheless able to detect its *effects*, and this allows him to conclude that magnetism exists. That is, if a piece of paper is placed over a magnet, and iron filings are sprinkled onto the paper, the filings jump into alignment, according to the magnetic field.

Magnetism itself is something man cannot experience by direct observation, but man can have experience of its *effects*, and thus he is able to reach the conclusion that the force of magnetism exists.

Similarly, it is true that man cannot experience the actual manufacture of anything alive. The design of a live object, however, is an effect that is no less real than the movement of iron filings. The effect must have a cause, even if the cause itself is something that man cannot experience directly.

The inconsistency is glaring: Seeing that people have no trouble using the iron filings to conclude that magnetism exists, why do they have so much trouble using design in nature to conclude that G-d exists? Seemingly, just as people are justified in inferring that magnetism is the force behind the movement of iron filings, people are equally justified in inferring that G-d is the Force behind the fabulous design in nature!

Just as the movement of the filings is the only thing needed to prove the existence of magnetism, so, too, the design in live objects is the only thing needed to prove conclusively the existence of G-d.

People do not require personal experience when the issue is magnetism (and they do not require it in everyday life), yet they *do* feel that personal experience is necessary when the issue is G-d. Obviously, a double standard is being applied, and there is only one explanation for it. Cognitive dissonance. This is why the overwhelming evidence of G-d from design in nature is virtually ignored.

As noted above, whether or not people actually *should* experience cognitive dissonance when thinking about G-d is not the issue here. Clearly, however, people do experience dissonance when thinking about G-d, and it is primarily because they have mistaken notions about G-d, are lacking information and have made large emotional and intellectual investments elsewhere.

In a sense, then, the dissonance is unjustified. Nevertheless, it exists and is quite strong. And because of it, people inflate the importance of personal experience in decision-making, even though in their normal lives, they make plenty of decisions without the

benefit of personal experience. Because of the cognitive dissonance, nature's overwhelming evidence of G-d is ignored, the
investments are protected, and the ignorance remains.

CHAPTER EIGHT

JUDAISM ON BELIEF IN G-D

I n the Five Books of Moses, the text states that the entire Jewish nation heard the voice of G-d at the giving of the Ten Commandments, and the first words G-d said to the Jews were, "I am G-d, your Lord, Who took you out of the land of Egypt."

According to the illustrious Maimonides, one of the great medieval codifiers of Jewish law, these opening words are not to be taken simply as a statement G-d made in order to identify Himself.

Rather, these opening words from G-d are Jewish law. They are a command, such as "You shall not steal" or "You shall honor your father and mother." Maimonides explains that this dictum is G-d's command to the Jewish people "to know and to believe in G-d," which is the very first commandment of the Ten Commandments.

Rabbi Elchonon Wasserman, a great twentieth-century sage,

questioned how Maimonides could offer such an explanation.[38] Rabbi Wasserman wrote that regarding actions, the concept of a "command" is understood. G-d can command the Jews to perform certain actions, such as "honor your father" or "keep the Sabbath."

Conversely, He also can command the Jews to *refrain* from certain actions, such as in "You shall not steal." For to do an act or not do an act is dependent on a person's will, and it is within a person's power to do or to refrain from doing. But how can Maimonides assert that G-d commands the Jews to *believe*? Belief is not an act.

Seemingly, it is not something a human being can control. Regarding a person's belief in G-d, it is not easy to see how the concept of a "command" is even pertinent!

Furthermore, if the explanation by Maimonides of the first of the Ten Commandments is correct, then even young boys aged thirteen would be obligated by that commandment to believe in G-d, because once a boy reaches thirteen, he becomes obligated to fulfill every one of the Ten Commandments.

But this aggravates the problem. For, as Rabbi Wasserman observed, belief in G-d has been a matter of great debate amongst intellectuals much older and much more advanced than normal boys of thirteen, yet many of these intellectuals have doubted G-d's existence or denied it completely. Seeing that intelligent, older people do not necessarily attain belief in G-d, how can Maimonides say that G-d commands Jewish boys who are only thirteen to attain this intellectual level and overcome all opposing arguments?

Rabbi Wasserman answers, defending the explanation of Maimonides as follows. In truth, nature and the human body provide so much testimony to G-d's existence that *even a thirteen-year-old should be able to conclude and be fully convinced intellectually that G-d exists*. In fact, as Rabbi Wasserman wrote, "G-d's existence should be obvious to anyone who has not taken leave of his senses." By virtue of the grand design in nature and in man's body, recognition of G-d should come to a person *intuitively*, without the person's having to think much about it at all.

According to Rabbi Wasserman, great intellects have doubted and denied G-d's existence only because they were victimized by internal personal pressures that confused and distorted their thinking and made them biased. The first of the Ten Commandments, Rabbi Wasserman therefore emphasizes, is actually G-d's command to free one's self from these internal pressures, so the exceedingly obvious truth about G-d can be perceived.

In other words, philosophers (such as those of the Myopic Man school) have argued against the idea of G-d, and they have doubted and denied G-d because each one had personal "investments" elsewhere, namely, a lifestyle, a mind set, a desire, a habit, a self-image or some other personal quirk he subconsciously knew was inconsistent with the idea of G-d, but which he was not willing to give up or change, no matter what the price.

When a person with such dearly held "investments" is faced with the evidence of G-d, the desire to hold onto the investments creates great cognitive dissonance in the person, and the result is that he will not believe, although the evidence of G-d is compelling.

According to Judaism, Rabbi Wasserman explained, G-d's existence is something obvious, even to a lad, and the greatest obstacle between a person and the truth about G-d is the person himself. "Investments" bias a person. To attain belief in G-d, deep research and logical, cognitive, step-by-step proofs are not necessary, because man can recognize G-d intuitively from nature or from the design of his own body. Actually, to attain belief in G-d, the only thing required is knowledge of self and an honest "scouring-out" of one's own biases. Unfortunately, many people, especially scientists and philosophers, cannot free themselves of bias, so the obvious truth is lost on them.

THE WISDOM OF SOLOMON

A second Jewish teaching needs to be mentioned as well, for it is very important for scientists, particularly those such as Robert Shapiro.

As you may recall, Shapiro has written a best-selling book, which reveals that all the theories of life's origin advocated in the halls of science today are really mythology, with no basis in fact. Very nicely, the book documents how scientists today do not have even one believable theory about how life on earth really began. The main problem of science is that even simple life is so complicated, no one knows how it possibly could have "snapped" into being.

Shapiro makes mention of the "old-time" theory that G-d is responsible, but this old "religious" theory is not acceptable to him, because in his view, it is not "scientific." A scientist, according to Shapiro, who allows G-d into the picture is going "beyond the bounds of science." Rather, in Shapiro's view, scientists must continue to seek out other theories of life's origin, namely, theories leaving out the factor of G-d. By definition, Shapiro feels, theories of life's origin which are "scientific" cannot acknowledge G-d, because the "theory" that G-d is responsible for the existence of life cannot be tested by "the scientific method."

Today, he feels, there is still some chance that scientists will come up with a reasonable G-dless theory of life's origin, and in the future, he writes, if they fail to formulate such a G-dless explanation it nevertheless would be proper for scientists "to attempt to sort out the surviving, less probable scientific explanations" rather than admit to G-d.

One can speculate as to why Shapiro wants so badly to keep G-d "out," but speculations aside, there are words from King Solomon that Shapiro and people like him would do well to read. In *Ecclesiastes*, King Solomon writes:

> And I beheld all of G-d's works and saw that no man will ever be able to decipher what has been made under the sun. Man shall struggle to seek it out but he will not find it. Though a wise man may say he knows, he will not be able to discover it.(8:17)

In another place, King Solomon writes similarly:

> G-d made everything beautiful in its time, and He also put the hidden into people's hearts. But man will not find what G-d made from beginning to end.(3:11)

King Solomon is saying that mankind can attain an appreciation for nature's harmony (the "hidden" is in our hearts, we see the beauty), and mankind can also appreciate aspects of the grand design, but in the end, the exact workings of things will never be understood by man, no matter how hard he struggles. G-d's design is much too complicated for the mind of man to fathom, and life, whose workings are a mystery, will remain a mystery, even after the existence of G-d is admitted.

How the first life originated is only one of many enigmas confronting man when he looks at nature. Other enigmas are the co-related string of "coincidences" allowing for the formation of atoms and molecules, and the sun and solar system, all of which is the indispensable inorganic "background" for the enigma of life.

According to King Solomon, many of the enigmas facing science today will remain enigmas until the end of time, because the Supreme Wisdom which is behind it all is far beyond the wisdom and grasp of all mankind. For a resident of G-d's universe, the many enigmas "come with the territory," as it were, like it or not.

In time, Shapiro is confident, science will find the answers for everything under the sun. Having "invested" in such a mind set, he will not let G-d into the picture. It says in *Ecclesiastes* that science cannot find all the answers, and according to Judaism, *Ecclesiastes* can be relied upon, because the entire work was divinely inspired. Accordingly, scientists such as Shapiro who have invested so deeply in the scientific method, and who place so much confidence in it, are doomed to careers of frustration, and the investment they have made, which is particular to scientists, literally blinds them to an obvious truth.

Seemingly, the healthier approach, and the one more connected to reality, is the approach of the scientists of the Anthropic School of Thought. These scientists acknowledge G-d, admit that there are certain enigmas that never will be solved, yet continue to apply the scientific method to nature, trying to decipher what they can.

BEYOND THE BOUNDS OF SCIENCE

In all fairness, however, we must admit that some of the criticism Shapiro levels at the Anthropic School is quite justified. After having conceded that G-d exists, and that the design in the universe is the work of G-d, scientists of the Anthropic School say that natural constants could not have emerged from chaos without the intervention of G-d. Likewise, once the inorganic realm was founded on the basis of these constants, the inorganic realm could not have brought forth life, even in its most simple form, without G-d. And certainly, human intelligence would never have emerged had it not been the Will of G-d.

G-d must have had the "anthropic" goal "in mind" from the very beginning. His original goal must have been to produce man, the creature with the mental and aesthetic faculties who can appreciate G-d's design. The universe was contrived only to produce man, and the purpose of man, to use a phrase from the 1960s, is to "dig it." So say the scientists who have "come over" to the side of G-d.

Here, Shapiro is absolutely right in saying that the Anthropic School of Thought has stepped far beyond the bounds of science. Since when are scientists qualified to be philosophers? Furthermore, scientists admit they do not know how things in nature work, and they certainly do not understand the workings of the entire universe. Given that scientists looking at the universe cannot fathom the design itself, how can they presume to know the design's purpose?

The matter can be likened to the following scenario. We are at the launching of a most sophisticated spaceship, designed by men possessing the greatest minds in the world, who have built it to perform a number of functions. As the great machine lifts off its pad, a newspaper reporter notices a child watching the liftoff amazed, with his mouth hanging open.

The reporter asks the child, "Why is the rocket going up into the sky? What is it supposed to do up there?"

The child, of course, hasn't the foggiest notion of how the ship is built or how it works, and neither does he know the purpose of the thing. As a result, his answer to the reporter is simple, "Why ask me? Ask the ones who built it!"

So, too, the world and the universe constitute the greatest "spaceship" of all. Every person is like the child, looking at the world and the universe with eyes of wonder, not knowing how the many components are put together, or how they work, or how they came into being in the first place. How can he be expected to know the purpose for it all?

If a person, such as a scientist from the Anthropic School, is able to rise above cognitive dissonance and realize the obvious fact that the design in the universe is G-d's design, he should have the sense to concede that G-d and the universe are much bigger than he, and *he should admit to himself that he is simply incapable of determining on his own the purpose for it all.* He should say to himself, "Only the Designer Himself knows the purpose."

This is what King Solomon was saying. Judaism teaches that it is G-d Who knows the purpose of creation, and the only way one can come to understand it is to listen to the Word of G-d, which was given to man in the Torah.

Judaism teaches that the entire universe is, in fact, created for man, and if in all the universe, there was only one lone human being, the universe would exist only for him. However, the reasons G-d created the universe for man, and what man's purpose here is, are things we can only learn from G-d Himself, by learning His Torah.

CHAPTER NINE

ODYSSEY

The discussion up to this point has focused primarily on why people are often not believers in G-d. Why people might have a hard time coming to accept the Torah is a separate subject entirely.

There are reasons a person might suffer cognitive dissonance when thinking about the Torah, even though again, the evidence of the truth of the Torah is compelling, as compelling as nature's evidence of G-d.

The ancient Greek philosopher Alexander Aphrodisius writes of three different factors that act as "blocks" to a person's seeing truth: arrogance, vainglory and the love of freedom; the subtlety, depth and difficulty of the subject at hand; human ignorance, want of sufficient mental capacity.

According to the Torah, as explained by Maimonides, a fourth factor stands in the way of perceiving the truth: habit and training.

In his *Moreh Nevuchim*, Maimonides explained that

> as human beings we naturally like what we have be-
> come accustomed to, and are attracted towards it. This
> may be observed among primitive rural villagers.
> Though they rarely enjoy the benefit of a bath, and they
> pass a life of privation having few enjoyments, still, they
> dislike town life and do not desire its pleasures,
> preferring inferior things to which they are accustomed
> over the better things to which they are strangers. It would
> give such rural villagers no pleasure to live in palaces, to
> be clothed in silk and to indulge in baths, ointments and
> perfumes.
> The same is the case with opinions of man to which he
> has been accustomed from his youth. He likes those opin-
> ions. He defends them and shuns any different views.
> This is yet a fourth factor which prevents a person from
> finding the truth.[39]

In other words, knowledge and understanding can be kept
from a person by his own subconscious, because a person
naturally prefers the ideas to which he is accustomed. The truth
can be lost on a person if in some way it does not "fit" with his
present way of thinking or with his habits and status quo. There
is a certain inertia that must be overcome, because where a person
has made "investments," he has a hard time giving them up.

Hopefully, the limited discussion here will serve as a stimulus
to prompt people to examine their "investments" and think more
deeply about both G-d and the Torah. Regarding the Torah,
certain programs are available which illustrate various unique
phenomena in the Torah text, which are themselves unique

elements of design. These programs also provide a more thorough treatment of the subject of G-d. Another excellent way to learn about G-d and the Torah is to attend a *yeshivah*, where G-d's Torah is studied in depth. At any rate, overcoming inertia is the first step.

In sum, for those who want to procure the truth about these very important issues, the suggested course of action is this: First, there must be a stage of introspection. One may have "invested" in certain ideas, attitudes and modes of behavior capable of stopping one from seeing the obvious. One needs to realize that the greatest obstacle between oneself and the truth may be one's own self.

One needs to discover one's own personal "investments," which might block out the truth about G-d and the Torah, and one cannot discover these investments except through one's own introspection and careful examination of one's own thoughts, feelings and ideas.

Second, there must be a stage of research into G-d and into the Torah, and the sources to be consulted must be reliable. Since a special relationship exists between G-d and the Jews, and the relationship arises from the Torah, the best sources to be consulted are the Jews of the Torah.

Last, the thinker needs freedom from external confusion. A certain level of detachment must be achieved. The thinker needs to realize that the opinions of the masses are not necessarily the result of logic and pure reason.

Not only scientists, but laymen as well, have invested heavily in the G-dless view of the universe, and some probably will never relinquish their precious investment. As long as they subconsciously regard G-d as some sort of lion about to pounce, they will not see the obvious truth, and they will not even want to think about the idea of G-d.

Much of what has been put forth here can be found in an article entitled "Science, G-d and Man," which appeared in *Time* magazine, the first issue of the year 1993. *Time* summarizes the recent trend in science—that all the evidence points to G-d. Some of the biggest names in science are quoted. One scientist has a book coming out that actually has G-d in its title. Another has a book mentioning G-d in the text, not in the title, but these scientist are like Robert Shapiro: They admit that G-d seems to be the likely explanation for everything that has been discovered, but it is "unscientific" to even mention G-d! Aware of the many mysteries of design that science has uncovered, all these scientists can say is that "we cannot peer behind the curtain." *Time* reports that these scientists are troubled, however, because "if you admit that we cannot peer behind the curtain, how can you be sure that there is nothing there?"

Nevertheless, Shapiro and the others continue in their ways. *Time* quotes William D. Hamiltion of Oxford University and calls him "one of the greatest scientific minds of our era." In the mind of Hamiltion, the idea that G-d is responsible "is certainly alive," but the seeding theory, too, is "a hypothesis that is very, very hard to dismiss." Hamiltion admits that seeding was stated "in a joking spirit," but still, it is hard to dismiss. Why? Because if it is dismissed, as it should be, all that's left is the "lion." Scientists, out of ignorance and irrational fear, will continue to look for ways to escape what is obvious, and the media will report it. In the meantime, reluctance of scientists to admit to G-d will itself serve as a block to the man on the street.

Time, however, to its credit, does write that someone who is "religiously inclined" and "wouldn't mind seeing some hard evidence" can find it today in science.

FREEDOM OF CHOICE

As the scientists of the Anthropic School say, man is a unique creature. All other creatures in nature operate exclusively under fixed, natural, immutable laws. A spider, a bird, a horse, or a monkey, each lives according to its drives, urges and instincts, and nothing more. Each has no choice but to live this way; they are "fixed" in nature, and completely subject to it.

Man, however, has the power to choose how he lives, and this power to choose makes him different from any animal. Freedom of choice is a sign that man has a soul. Yes, man does have a "horse" in him, an "animal" side, and this "horse" is in fact pulled in the direction of its urges. Man, however, the creature with choice, has another system which can intervene and guide the horse in a completely new direction. Every human being has the ability to choose what is good, true and moral, despite his urges. This makes man different. It allows him to operate *above* the laws of nature.

Man's relationship to truth sets him apart from the animals. To a human being, the value and importance of truth is self understood, but monkeys and horses do not even have the concept of truth. Even if an animal could be taught to speak, he could not understand the concept of truth, and certainly he could not grasp its importance. Only man has the concept and the appreciation, and only man has the desire for truth.

To attain truth and to live a life of truth, is actually to connect to reality. Such a connection should be sought by everyone, for what good is a life of falsity or half-truth? Everyone, even the intelligent and the righteous, experiences cognitive dissonance, regarding many different things. Cognitive dissonance can block a person from seeing the truth, so when the issue is G-d and the stakes are high, one must especially be on the lookout.

Now, aware of "cognitive dissonance," the most important truth that exists is much more accessible to you. You do not need to be a scientist, or a person of superior intelligence, to be able to see G-d in nature. Because nature's design makes G-d's existence obvious, even a thirteen-year-old has this ability. The Stephen Jay Goulds, the Robert Shapiros, the Cricks and Orgels of the world and the philosophers of the Myopic Man school all missed the obvious truth, in spite of their intelligence, and the only reason they missed it is that they were kept from it by their own biases.

Cognitive dissonance can prevent a person from accepting constructive criticism. Cognitive dissonance can prevent a person from recognizing a mistake he made while shopping. But cognitive dissonance also can stop a person from recognizing G-d, and the result in this case is that one misses out enjoying life as a being in touch with the real world. It behooves the intelligent person to open up to G-d and get to know Him through His Torah.

FOOTNOTES

1. Einstein: 1879-1979, *Jewish National and University Library*, *March 1979*. Jerusalem. p. 10. (a catalogue published in conjunction with the Organizing Committee of the Jerusalem Einstein Centennial Symposium, organized by the Israel Academy of Sciences and Humanities, The Hebrew University of Jerusalem, TheVan Leer Jerusalem Foundation, The Jerusalem Foundation and the Aspen Institute for Humanistic Studies, Jerusalem.)
2. Ibid. p.49.
3. Beyond the Milky Way, *Time-Life Inc.* New York. p. 148.
4. Einstein: 1879-1979. p. 16.
5. Ibid. p. 47.
6. Ibid. p. 20.
7. Ibid. p. 10.
8. Beyond the Milky Way. p. 178.
9. New York Times Magazine, June 25, 1978.
10. Ibid.
11. Time Magazine, February 5, 1979.
12. *From Conception to Birth*. Harper and Row Publishers. New York. 1970. Pp. 34, 57-58, 76.

13. *The Disputation*. Scholarly Publications. England. 1979. p. 178.

14. Dietrick E. Thompson. The Universe, Chaotic or Bioselective? *Science News, Vol. 106, October 19.* 1974. p. 124. (quoting Dr. Freman J. Dysan of The Institute for Advanced Study, Princeton, New Jersey.)

15. Davies, Paul. *The Accidental Universe.* Cambridge University Press. 1982. p. 7.

16. Ferre, Frederick. *Basic Modern Philosophy of Religion.* Charles Scribner's Sons. New York. 1967. p. 161.

17. Garardin, Lucien. *Bionics.* Translated from the French by Pat Priban. World University Library. London. 1968. p. 41.

18. Klahr, C. N. Science Versus Scientism, *Challenge, Torah Views on Science and Its Problems.* Carmel, A. and Domb, C. (Ed.). Feldheim Publishers and the Association of Orthodox Jewish Scientists. Jerusalem, New York, and London. 1978. p. 291.

19. Koestler, A. and Smythies, J. R. (Eds.). *Beyond Reductionism—New Perspectives in Life Sciences.* Hutchinson and Co. London. 1969. (The Alpach Symposium.) p. 66.

20. Hoyle, Fred and Wickramasinghe, Chandra. *Evolution from Space.* J. M. Dent and Sons Co. London. 1981. p. 148.

21. Was Darwin Wrong? *Life Magazine, April, 1982.* Excerpted from *The Neck of the Giraffe*, by Francis Hitching, Ticknor and Fields, 1982.

22. *Evolution from Space.* p. 147.

23. Ibid. p. 148.

24. Ibid. p. 131.

25. Ibid. p. 24.

26. Time Magazine, November 21, 1983. p. 49.

27. Shapiro. *Origins: A Skeptic's Guide to the Creation of Life On Earth.* Bantam Books. New York. 1986. p. 98-116.

28. Ibid. p. 99.

29. Ibid. p. 109.

30. *Evolution from Space.* p. 31.

31. Ibid. p. 148.

32. Ibid. p. 30.

33. *Origins.* p. 227-8.

34. Ibid. p. 263.

35. *Evolution from Space.* p. 31.

36. Ibid. p. 137.
37. Sagan, Carl and Shklovski, I. S. *Intelligent Life in the Universe*. Dell Publishing Co. New York. 1966. p. 252.
38. Wasserman, Rabbi Elchonon. *Kovetz Maamarim, Chapter One.*
39. Maimonides. *Moreh Nevuchim.*

SUGGESTED READING

TORAH AND JUDAISM

The Living Torah, Rabbi Aryeh Kaplan (Moznaim)
One of the most readable translations of the Five Books of Moses. Includes footnotes, maps, tables, charts, bibliography and index.

The Eternal Link, Rabbi Pinchas Winston (C.I.S. Publishers)
A brief synopsis of the biblical period and the contents of the twenty-four books of the *Torah*. Includes foldout, full color comparative timelines.

To Be A Jew, Rabbi Chaim Donin (Basic Books)
This complete guide to Jewish observance in contemporary life clearly explains every holiday and ritual.

The Book of Our Heritage, Rabbi Eliyahu Kitov (Feldheim)
A thorough, month-by-month explanation of the different customs and laws of the entire Jewish year.

DEALING WITH THE ISSUES

Anti-Semitism: Why The Jews? Prager-Telushkin (Simon & Schuster)
This thorough and analytical study of the root of anti-Semitism leaves the reader with an appreciation of Judaism and Jewish history.

The Jew in Exile, Rabbi Shmuel Yitzchack Klein (C.I.S. Publishers)
An insightful review of the dynamics of anti-Semitism and role of the Jew in the modern world.

Being Jewish, Rabbi Shimon Hurwitz (Feldheim)
A sharply critical analysis of Western society contrasted with an overview of Jewish values.

If You Were G-d, Rabbi Aryeh Kaplan (NCSY)
These three short essays will really get you thinking about the nature of G-d, existence, and life after life.

Permission to Believe, Laurence Keleman (Feldheim)
The author presents a rational proof for G-d's existence using four separate intellectual approaches, dispelling the misconception that belief in G-d is irrational.

Challenge, compiled by Rabbi Karmel & Domb (Feldheim)
Torah views on recent scientific advances include studies by various authors presenting Torah solutions to ethical dilemmas.

Genesis and the Big Bang, Gerald I. Schroeder (Bantam)
Discovering harmony between modern science and the Bible. This groundbreaking book shows that the apparent gap between the theories of both is actually quite narrow.

ETHICAL BEHAVIOR AND CHARACTER DEVELOPMENT

Strive for Truth, Rabbi Eliyahu Dessler (Feldheim)
One of the leading *mussar* (character development) works of this century, gaining human nature through the teachings and wisdom of the Torah.

Beyond Your Ego, Dr. Judith Mishell & Dr. Shalom Srebrenik (C.I.S. Publishers)
A penetrating examination of the psychological and emotional aspects of the human mind and soul.

Let Us Make Man, Dr. Abraham J. Twerski (C.I.S. Publishers)
The nationally acclaimed psychologist plumbs the wisdom of the *Torah* to develop guidelines for the attainment of self-esteem through Jewishness.

PERSONAL EXPERIENCES

Generation to Generation, Dr. Abraham J. Twerski (C.I.S. Publishers)
Dr. Twerski, who is a scion of an illustrious *chassidic* family, describes life in his father's home in Milwaukee. A classic.

The View from Above, Rachel Noam (Bristol, Rhein & Englander)
A *kibbutz* woman has an out-of-body, near-death experience (NDE) and the revelation eventually steers her to observe Judaism.

The Holocaust Diaries, (C.I.S. Publishers)
A collection of memoirs describing the devotion of *Torah* Jews to their people and their faith. Five volumes: Late Shadows, They Called Me Frau Anna, Dare to Survive, Behind the Ice Curtain, Sisters in the Storm.

Deep in the Russian Night, Aaron Chazan (C.I.S. Publishers)
The story of a family that refused to compromise on its *Torah* observance through the darkest periods of Stalinist oppression.